Family Business and Technological Innovation

Alessandra Perri · Enzo Peruffo

Family Business and Technological Innovation

Empirical Insights from the Italian Pharmaceutical Industry

Alessandra Perri
Department of Management
Ca' Foscari University
Venice, Italy

Enzo Peruffo
Department of Business and Management
LUISS Guido Carli
Rome, Italy

ISBN 978-3-319-61595-0   ISBN 978-3-319-61596-7   (eBook)
DOI 10.1007/978-3-319-61596-7

Library of Congress Control Number: 2017947743

Cover illustration: © nemesis2207/Fotolia.co.uk

Printed on acid-free paper

This Palgrave Macmillan imprint is published by Springer Nature
The registered company is Springer International Publishing AG
The registered company address is: Gewerbestrasse 11, 6330 Cham, Switzerland

"To the memory of Nonna Rita"
Alessandra

"To my son Andrea"
Enzo

Acknowledgements

The authors wish to acknowledge the financial support of the LUISS Business School and the Department of Management of Ca' Foscari University of Venice.

Contents

Abbreviations, Acronyms, or Chronology

BAM	Behavioral Agency Model
Big Pharma	Big Pharmaceutical Companies
CEO	Chief Executive Officer
GDP	Gross Domestic Product
R&D	Research and Development
RBV	Resource-based View
ROA	Return on Assets
SEW	Socio-Emotional Wealth
SMEs	Small and Medium-sized Enterprises
SSN	Sistema Sanitario Nazionale
USA	United States of America
USPTO	US Patent and Trademark Office

List of Figures

List of Tables

1

Introduction

Abstract This introduction offers a synopsis of the main topics and arguments covered in this volume. The book is divided into two parts. The first part (Chaps. 2, 3) offers a general perspective on the dynamics of innovation in family firms, by reviewing the most relevant theories that, from both a family business and an innovation studies viewpoint, contribute to understand the management of technological innovation in family firms. The second part (Chaps. 4, 5) works with selected theories to carry out an analysis of family firms' innovation performance in the Italian pharmaceutical industry. It describes the industry setting and the empirical methodology and discusses the results of the analysis. The final Chapter (Chap. 6) concludes and offers directions for future research.

Keywords Chapters' overview · Family firms · Innovation Pharmaceutical industry

© The Author(s) 2017
A. Perri and E. Peruffo, *Family Business and Technological Innovation*,
DOI 10.1007/978-3-319-61596-7_1

1.1 Introduction to Family Business and Technological Innovation

The relationship between family business and innovation is unique in many respects (Duran et al. 2016; De Massis et al. 2015. Family firms are typically depicted as conservative, risk-averse and path-dependent organizations, with limited willingness to embrace change and to invest in innovative activities (Gómez-Mejía et al. 2007; Economist 2009). At the same time, many of the most successful, enduring, and innovative companies worldwide are indeed family firms (Forbes 2014; Duran et al. 2016; Family Firm Institute 2017).

This puzzling phenomenon has recently attracted widespread attention, as scholars have tried to explain how different dimensions of a family's involvement in a firm's business may influence the firm behavior in terms of innovation inputs (Block 2012; Gómez-Mejía et al. 2014; Matzler et al. 2015; Schmid et al. 2014; Duran et al. 2016) and outputs (Block et al. 2013; Konig et al. 2013; Carnes and Ireland 2013; Matzler et al. 2015; Duran et al. 2016; Cucculelli et al. 2016; Kammerlander et al. 2015; De Massis et al. 2016).

While innovation scholars have extensively investigated the most relevant drivers and contingencies explaining firm innovation, the understanding of the role of family firms in the management of technological innovation is still limited. This can be at least partially explained by the fact that corporate governance institutions and mechanisms have started to be explicitly considered as possible influencing factors of a firm's innovation decisions and performance only in recent decades (Bushee 1998; Coriat and Weinstein 2002; Hall and Soskice 2001; Lee and O'Neil 2003; O'Sullivan 2000; Tylecote and Ramirez 2006). Yet, several theoretical and empirical hints suggest that innovation decisions, processes, and outcomes differ across family and non-family firms (De Massis et al. 2013). For example, previous literature has argued that family involvement in a firm's ownership, management, and governance could help to develop unique resources that nurture technological innovation (e.g., Sirmon and Hitt 2003). At the same time, empirical research has demonstrated that to protect their power, control, and legitimacy, family firms may exhibit lower propensity to acquire external technology compared to non-family firms (Kotlar et al. 2013). In general, the documented differences in innovative behavior across family and non-family firms, coupled with the recognition that the former are the most widespread type of firm governance (Chrisman et al. 2015), explain why scholars are paying increasing attention to family firms innovation.

From a theoretical viewpoint, this nascent strand of literature has mainly leveraged family business research, ranging from traditional agency (Matzler et al. 2015; Schmidt et al. 2014; Duran et al. 2016) and resource-based (Sirmon and Hitt 2003; Sirmon et al. 2008) theories, to stewardship theory (Davis et al. 2010), socio-emotional perspectives (Gómez-Mejía et al. 2007), and behavioral agency model (Wiseman and Gómez-Mejía 1998). While a prevailing reliance on family business research is natural for a literature stream that is rooted in the interest and curiosity of family business scholars, this book aims at complementing this viewpoint with a more systematic use of constructs and tools arising from the innovation studies literature. Strengthening both theoretical building blocks, this book seeks to provide an integrative framework for investigating technological innovation in family firms. Understanding the multifaceted choices, influences, and implications related to the management of innovation in family firms requires accounting for the role of various governance, institutional, and technological factors. In turn, distinct theoretical frameworks and empirical instruments are necessary to master the phenomenon of analysis. This book thus works with a balanced combination of selected family business and innovation studies literature, with the aim of contributing fresh insights into existing studies of innovation in family firms.

Empirically, this book investigates a very intriguing setting, i.e., the pharmaceutical industry, with a focus on the Italian family firms that compete in this global arena. Existing empirical studies on innovation in family firms have mainly focused on contexts in which technological innovation is not a core driver of firm performance (Gómez-Mejía et al. 2014). In this study setting, two industry conditions make the family–innovation puzzle much more pronounced and, hence, interesting to explore: first, the marked and above average technology intensity of the pharmaceutical industry, which requires companies to invest huge amounts of resources in innovative activities that are increasingly risky and uncertain (Bruche 2012), clashes even more sharply with family firms' typical conservative approach and risk aversion (Gómez-Mejía et al. 2007); second, the inherently global nature of the pharmaceutical industry, besides conflicting with family firms' traditional focus on local demand and factor markets (Fernández and Nieto 2005), further increases their competitive pressure, thus encouraging to pursue continuous technological upgrading by means of innovative activities. Therefore, in this more than in other contexts, family-driven incentives seem to conflict with industry-driven incentives.

From a methodological viewpoint, the project follows the tradition of existing studies that have mainly leveraged quantitative approaches (De Massis et al. 2013) but also employs qualitative data as a complement to gain more profound insights into the governance and innovation dynamics animating this peculiar empirical setting. More specifically, this study uses a blended approach that combines a core of quantitative analysis conducted over a sample of Italian pharmaceutical companies, with qualitative evidence collected through face-to-face interviews with both managers from a subset of our sample firms and other industry experts (see Scalera et al. 2014 for a similar approach). Through this multimethod research design, this study allows to:

- Overcome the focus on single dimensions of innovation output, to explore a broader and more comprehensive set of multifaceted constructs that enable to describe family firms' innovative behavior in a more accurate way;
- Reconstruct the links between different dimensions of family involvement and the firm's innovative performance in a particular institutional context (i.e., Italy) wherein family firms represent a truly pervasive phenomenon.

In terms of structure, the book is composed of two main parts. The first part (Chaps. 2, 3) offers a general perspective on the nature and the dynamics of innovation in family firms and reviews the most relevant theories, and resulting empirical evidences, that enable to account for the strategic and organizational specificities characterizing the management of technological innovation in family firms. The second part (Chaps. 4, 5) describes the industry setting and the empirical methodology and discusses the findings of the analysis. For the sample firms considered in this study, a set of corporate governance dimensions are explored in association with a number of aspects qualifying the performance of firm innovative activities. Finally, Chap. 6 concludes by proposing a framework for the analysis of innovation in family firms competing in high technology and global settings, a range of best practices and an updated research agenda to inform future studies.

On the whole, the study seeks to uncover unique mechanisms linking the ownership, management, and governance dimensions of the family to different performance facets of technological innovation. Hence, it contributes to understand the idiosyncratic aspects of family businesses that may influence how innovation is managed and generated in this organization context, and their implications in terms of innovative outcomes. In doing so, it covers the most important theoretical perspectives required to grasp a complex phenomenon such as innovating in conservative, path-dependent, and discretionary organizations. Bridging the well-established streams of literature on family firms with an accurate account of the specificities of technological innovation management, it tries to address the need for a focused, timely yet theoretically and empirically robust discussion of this relevant phenomenon.

References

Block, J. H. (2012). R&D investments in family and founder firms: An agency perspective. *Journal of Business Venturing, 27*(2), 248–265.

Block, J., Miller, D., Jaskiewicz, P., & Spiegel, F. (2013). Economic and technological importance of innovations in large family and founder firms an analysis of patent data. *Family Business Review, 26*(2), 180–199.

Bruche, G. (2012). Emerging Indian pharma multinationals: Latecomer catch-up strategies in a globalised high-tech industry. *European Journal of International Management, 6*(3), 300–322.

Bushee, B. J. (1998). The influence of institutional investors on myopic R&D investment behavior. *Accounting Review, 73*(3), 305–333.

Carnes, C. M., & Ireland, R. D. (2013). Familiness and innovation: Resource bundling as the missing link. *Entrepreneurship Theory and Practice, 37*(6), 1399–1419.

Chrisman, J. J., Chua, J. H., De Massis, A., Frattini, F., & Wright, M. (2015). The ability and willingness paradox in family firm innovation. *Journal of Product Innovation Management, 32*(3), 310–318.

Coriat, B., & Weinstein, O. (2002). Organizations, firms and institutions in the generation of innovation. *Research Policy, 31*(2), 273–290.

Cucculelli, M., Le Breton-Miller, I., & Miller, D. (2016). Product innovation, firm renewal and family governance. *Journal of Family Business Strategy, 7*(2), 63–130.

Davis, J. H., Allen, M. R., & Hayes, H. D. (2010). Is blood thicker than water? A study of stewardship perceptions in family business. *Entrepreneurship Theory and Practice, 34,* 1093–1116.

De Massis, A., Frattini, F., & Lichtenthaler, U. (2013). Research on technological innovation in family firms: Present debates and future directions. *Family Business Review, 26*(1), 1–22.

De Massis, A., Di Minin, A., & Frattini, F. (2015). Family-Driven Innovation. *California Management Review, 58*(1), 5–19.

De Massis, A., Frattini, F., Kotlar, J., Petruzzelli, A. M., & Wright, M. (2016). Innovation through tradition: Lessons from innovative family businesses and directions for future research. *The Academy of Management Perspectives, 30*(1), 93–116.

Duran, P., Kammerlander, N., Van Essen, M., & Zellweger, T. (2016). Doing more with less: Innovation input and output in family firms. *Academy of Management Journal, 59*(4), 1224–1264.

Economist. (2009). *Dynastie and durability.* http://www.economist.com/node/14517406.

Family Firms Institute. (2017). *Family enterprise statistics from around the world.* Available at: http://www.ffi.org/?page=GlobalDataPoints.

Fernández, Z., & Nieto, M. J. (2005). Internationalization strategy of small and medium-sized family businesses: Some influential factors. *Family Business Review, 18*(1), 77–89.

Forbes. (2014). *The World's Most Innovative Companies.* www.forbes.com/inno-vative-companies/list/#page:1_sort:0_direction:asc_search:_filter:Europe_filter:All%20industries.

Gómez-Mejía, L. R., Haynes, K. T., Núñez-Nickel, M., Jacobson, K. J., & Moyano-Fuentes, J. (2007). Socioemotional wealth and business risks in family-controlled firms: Evidence from Spanish olive oil mills. *Administrative Science Quarterly, 52*(1), 106–137.

Gómez-Mejía, L. R., Campbell, J. T., Martin, G., Hoskisson, R. E., Makri, M., & Sirmon, D. G. (2014). Socioemotional wealth as a mixed gamble: Revisiting family firm R&D investments with the behavioral agency model. *Entrepreneurship Theory and Practice, 38*(6), 1351–1374.

Hall, P. A., & Soskice, D. (Eds.). (2001). *Varieties of capitalism: The institutional foundations of comparative advantage.* Oxford: Oxford University Press.

Kammerlander, N., Dessì, C., Bird, M., Floris, M., & Murru, A. (2015). The impact of shared stories on family firm innovation a multicase study. *Family Business Review, 28*(4), 332–354.

König, A., Kammerlander, N., & Enders, A. (2013). The family innovator's dilemma: How family influence affects the adoption of discontinuous technologies by incumbent firms. *Academy of Management Review, 38*(3), 418–441.

Kotlar, J., De Massis, A., Frattini, F., Bianchi, M., & Fang, H. Q. (2013). Technology acquisition in family and nonfamily firms: A longitudinal analysis of Spanish manufacturing firms. *Journal of Product Innovation Management, 30*(6), 1073–1088.

Lee, P. M., & O'Neill, H. M. (2003). Ownership structures and R&D investments of US and Japanese firms: Agency and stewardship perspectives. *Academy of Management Journal, 46*(2), 212–225.

Matzler, K., Veider, V., Hautz, J., & Stadler, C. (2015). The impact of family ownership, management, and governance on innovation. *Journal of Product Innovation Management, 32*(3), 319–333.

O'Sullivan, M. (2000). The innovative enterprise and corporate governance. *Cambridge Journal of Economics, 24*, 393–416.

Scalera, G. V., Mukherjee, D., Perri, A., & Mudambi, R. (2014). A longitudinal study of MNE innovation: The case of Goodyear. *Multinational Business Review, 22*(3), 270–293.

Schmid, T., Achleitner, A. K., Ampenberger, M., & Kaserer, C. (2014). Family firms and R&D behavior—New evidence from a large-scale survey. *Research Policy, 43*(1), 233–244.

Sirmon, D. G., & Hitt, M. A. (2003). Managing resources: Linking unique resources, management, and wealth creation in family firms. *Entrepreneurship Theory and Practice, 27*(4), 339–358.

Sirmon, D. G., Arregle, J. L., Hitt, M. A., & Webb, J. W. (2008). The role of family influence in firms' strategic responses to threat of imitation. *Entrepreneurship Theory and Practice, 32*(6), 979–998.

Tylecote, A., & Ramirez, P. (2006). Corporate governance and innovation: The UK compared with the US and 'insider' economies. *Research Policy, 35*(1), 160–180.

Wiseman, R. M., & Gómez-Mejía, L. R. (1998). A behavioral agency model of managerial risk taking. *Academy of Management Journal, 23*(1), 133–153.

2

Theoretical Perspectives on Family Firms

Abstract This chapter offers a general perspective on family firms—the most widespread form of business organization. It begins with an analysis of various possible definitions of "family firm," distinguishing two main approaches: components-of-involvement and essence. This assessment establishes a challenge to traditional views that treat family firms as homogeneous. Next, a review of existing literature provides a systematic summary of the primary theoretical approaches adopted to investigate family firms: agency theory, the resource-based view of the firm, stewardship theory, and the behavioral agency model. This chapter concludes with a discussion of how the behaviors, goals, and interests of family firm owners define their firms' strategic decisions and performance.

Keywords Family firms · Family firm heterogeneity · Family business theory

© The Author(s) 2017
E. Peruffo, *Family Business and Technological Innovation*,
DOI 10.1007/978-3-319-61596-7_2

2.1   Family Firms: Definitions

Globally, family firms are the most common, diffused, and widely studied ownership structure, such that they contribute significantly not only to business and society as an essential business organization (De Massis et al. 2015) but also to organizational research as an empirical and theoretical research topic (Botero et al. 2015; De Massis et al. 2015; Carney et al. 2015). A recent business press report (The Economist 2015) indicates that more than 90% of the world's businesses are family managed or controlled, with varied influences across many different countries (e.g., Klein 2000; Anderson and Reeb 2003a; Morck and Yeung 2004; Villalonga and Amit 2006; Astrachan and Shanker 2003). For example, family businesses make up more than 80% of private-sector firms in the USA, employing 57% of the US workforce and contributing 63% of its gross domestic product (GDP) (McKinsey & Co. 2014; De Massis et al. 2015). Of the 500 largest family firms, 23.4% are in North America and account for 11.4% of North America's GDP; 46.4% settled in Europe, accounting for the 14.8% of the continent's GDP; and the rest are distributed across the Asia-Pacific and Latin America (Global Family Business Index, University of St. Gallen, Center for Family Business, EY Family Business Yearbook 2016). Research by Klein (2000) reveals that 58% of German and 71% of Spanish companies with more than €1 million in annual turnover are family businesses. In Europe, Italy is the nation with the greatest concentration of family firms; they account for 94% of its GDP (McKinsey & Co. 2014). This status likely has arisen because Italy offers several characteristics that are well suited to the emergence of a typical family business (e.g., Minichilli et al. 2010; Prencipe et al. 2011; Ling and Kellermanns 2010). Thus, Italian firms also exhibit a greater involvement of family members in key management positions (55% of family-controlled companies on the Milan Stock Exchange feature a family member as CEO; Minichilli et al. 2010).

With a three-cycle model, Lansberg (1988) denotes family membership, ownership, and management as the key features that distinguish family from non-family firms. Several subsequent attempts also seek to consolidate a clear definition of what constitutes a family firm (Chua et al. 1999; Sharma 2004; Pindado and Requejo 2015; Carney et al. 2013). For example, Chua et al.'s (1999) review of family firm definitions highlights the need to integrate essential elements of family business, such that they propose the following inclusive explanation:

> The family business is a business governed and/or managed with the intention to shape and pursue the vision of the business held by a dominant coalition controlled by members of the same family or a small number of families in a manner that is potentially sustainable across generations of the family or families. (Chua et al. 1999, p. 25)

Similarly, Astrachan et al. (2002) emphasize "soft" factors and develop a continuous F-PEC (family, power, experience, culture) scale to classify the family's influence. Anderson and Reeb (2003a) instead cite operational criteria to investigate differences between family and non-family firms in their financial performance.

The widespread diffusion of family firms, and the varied theoretical and empirical perspectives through which they have been investigated, thus leaves the definition of a family business unclear (Chua et al. 1999; Klein 2000; Astrachan et al. 2002; Sharma 2004; Pindado and Requejo 2015). The fragmentation also may be due to the operational nature of most research, such that scholars actively seek more extensive or tighter definitions, depending on their theoretical perspective and empirical setting (Klein 2000; Pindado and Requejo 2015). Thus, the same data set seemingly could lead to disparate results, depending on the definition used to classify the businesses as family-owned or not (Shanker and Astrachan 1996; Klein 2000).

A broad approach to account for the variety of family firm typologies would acknowledge that a family business is influenced substantially by one or more families in making its strategic choices (Shanker and Astrachan 1996; Klein 2000; Carney 2005; Fiegener 2010; Pindado and Requejo 2015). The influence stems from three dimensions: family involvement in company management, control, or family ownership (Carney 2005; Fiegener 2010; Pindado and Requejo 2015). In a comprehensive, empirical research review, Pindado and Requejo (2015) note that 57% of family firm studies use a definition associated with the ownership structure and 22% rely on a definition pertaining to family management or control. Thus, extant definitions tend to be operational in nature, without a systematic sense of which components (ownership, management, control) are most pertinent and in which proportions (Chua et al. 1999; Chrisman et al. 2005b).

Furthermore, the definitions adopted often reflect the nature of the study being conducted. Finance researchers generally employ an ownership structure definition; management scholars tend to focus on managerial or control elements. The results of such studies also reflect the empirical setting, and in this sense, financial scholars frequently analyze large, publicly listed family firms, with their extensive and easily accessible data (Classensen and Tzioumis 2006; Le Breton-Miller and Miller 2009; Pindado and Requejo 2015), while management scholars focus more on small to medium-sized, privately held businesses. Management studies accordingly suffer from a lack of readily available financial and ownership data, compared with the larger body of research on publicly listed companies, so they require softer, qualitative criteria to define family businesses (Klein 2000; Carney et al. 2013; Pindado and Raquejo 2015).

Among the confusion though, two theoretical approaches to defining family firms are widely accepted (Siebels and Knyphausen-Aufseß 2012). First, the *components-of-involvement approach* defines family firms according to the percentage of shares (i.e., decision-making rights) controlled by the same family or owner (Chua et al. 1999; Siebels and Knyphausen-Aufseß 2012; Schmid et al. 2015). The threshold changes, depending on the country or geographical area analyzed, to reflect differences in local institutional environments, country legal origins, institutional regulations, investor protection policies, stock markets, and overall rule-of-law standards (La Porta et al. 1998; Carney et al. 2013; Schmid et al. 2015). For example, in the USA, listed companies are defined as family firms if 5% of the voting rights concentrate with the same owner, but in the EU, the threshold is 25%, and the country-specific requirements range from 20% in France to 50% in Italy (Anderson and Reeb 2003a; Villalonga and Amit 2006; Minichilli et al. 2010; Prencipe et al. 2011; Schmid et al. 2015). This lack of consensus is the main limitation of the components-of-involvement approach; it leaves substantial room for interpretation and conflicting results.

Second, according to an *essence approach* to defining family firms, family involvement (i.e., ownership, management, and/or control) is a necessary and implicit but not sufficient condition to identify a family business (Chrisman et al. 2003; Chua et al. 1999; Habbershon et al. 2003; Siebels and Knyphausen-Aufseß 2012). The familial nature of a company thus depends on the behaviors of its members, which are distinctive with respect to those of non-family firms. These behaviors might include a willingness to influence the firm's strategic direction or vision, family involvement, social capital and emotional attachment, the pursuit of noneconomic values, and the adoption of a longer time horizon perspective (Chrisman et al. 2003; Chua et al. 1999; Habbershon et al. 2003; Gomez-Mejia et al. 2007). Compared with the more operational components-of-involvement approach, the essence view is theoretical in nature, allowing for the development of frameworks that specifically address family firms' distinguishing features. Yet it also lacks precise thresholds.

Thus, the best option may be the synergic adoption of both approaches. Starting with such a convergence objective, along with the set of definitional issues, several studies have sought to categorize different family business forms, in an effort to enhance understanding of family firm heterogeneity and shed light on existing theoretical approaches (Chrisman et al. 2005b; Westhead and Howorth 2006; Bammens et al. 2011).

2.2   Family Firm Heterogeneity

Melin and Nordqvist (2007) caution that ignoring family firm heterogeneity could lead to inaccurate understanding (Chua et al. 2012), and in response, family firm scholars try to distinguish not just family versus non-family firms but also categories within the set of family businesses (e.g., Melin and Nordqvist 2007; Chua et al. 2012; Pindando and Requejo 2015; Schmid et al. 2015). This heterogeneous group of firms contains differences that can influence key firm features, such as diversification decisions (Schimd et al. 2015), innovation (De Massis et al. 2015), and firm performance (Pindando and Requejo 2015). Accordingly, theoretical developments also consider the sources of family firm heterogeneity. Chua et al. (2012) recommend a categorization based on three main features of family firms: family goals (Chrisman et al. 2012; De Massis et al. 2015), resources (Habbershon et al. 2003), and governance structures (Carney 2005).

Owners of family firms tend to be concerned about both economic and noneconomic goals, and their levels of relevance can explain family firm heterogeneity. For example, the main element of a family firm is that it likely seeks to ensure family control and survival, because the firm functions like a family asset that can be passed on to the next generation. In some cases, this goal even overrides traditional profit maximization or value creation goals (Gomez-Mejia et al. 2007). Family involvement, in terms of ownership and management, thus is positively associated with the adoption of noneconomic goals, and family essence partially mediates this relation (Chrisman et al. 2012). In other words, the manner and extent to which the family influences firm decisions are a central driver of family firm heterogeneity. Some family firms actively pursue noneconomic outcomes such as family harmony or social status (e.g., De Massis et al. 2015), but others are more oriented toward profit maximization and wealth creation.

Adopting a resource-based view, which notes the roles of resources and capabilities in building competitive advantages (Barney 1991), family firm heterogeneity also might stem from the resources and capabilities that family owners require to reach their goals. Family business scholars show that path dependence in resource accumulation (Arregle et al. 2012), tacit knowledge and social capital (Lichtenthaler and Muethel 2012), and human capital (Sirmon and Hitt 2003; Verbeke and Kano 2012) all can be sources of family firm heterogeneity, in that they lead to differences in firm behavior and performance. For example, Verbeke and Kano (2012) identify a bifurcation bias—that is, the tendency of family firms to consider family managers as stewards but non-family managers as opportunistic agents—as a potential source of competitive disadvantage for large firms in high-tech industries. The relevance of this bifurcation bias depends largely on how the firm manages its economic and noneconomic goals, as well as the levels of trust and institutional development.

Finally, the governance structures adopted by family firms differ from those of non-family firms, reflecting distinct alignments of ownership, control, and management (Carney 2005). The heterogeneous roles of family owners, in terms of their varying involvement in ownership, control, and management, also might clarify heterogeneity at the firm level (e.g., Nordqvist et al. 2014; Schmid et al. 2015; Arregle et al. 2012). For example, if family members control majority ownership and are involved both in management and on the board (Sirmon et al. 2008), these owners have discretionary power over the firm's strategic options. They can leverage their social capital and indisputable control (Carney 2005) to gain advantages. Yet they also might suffer potential disadvantages, such as a greater risk of free-riding or redundant information (Arregle et al. 2012). Alternatively, if firms feature a strong family influence but not majority ownership, these family members have a less dominance over strategic decision making (Sirmon et al. 2008). In this case, powerful stakeholders (e.g., shareholders, directors, board members) may limit the ability of family owners to operate solely at their own discretion.

2.3   Theoretical Approaches to Family Firms

Reflecting this heterogeneity, multiple theories and frameworks have sought to disentangle the complexity surrounding family businesses. Scholars mostly rely on four pertinent theories: agency theory, the resource-based view, stewardship theory, and behavioral agency theory (Le Breton-Miller et al. 2015; Siebels and Knyphausen-Aufseß 2012; Bammens et al. 2011; Berrone et al. 2012).

2.3.1 Agency Theory

Agency theory is perhaps the most widely acknowledged theoretical approach to family firm behaviors (Jensen and Meckling 1976). Traditional agency theory anticipates opportunistic behaviors: An agent in a contract can operate in his or her interest rather than the interest of the principal, thus generating moral hazard and adverse selection problems (Eisenhardt 1989; Jensen and Meckling 1976). Traditional agency costs, which result from the so-called principal–agent problem (or Type I agency problem), arise when ownership and management incentives are not aligned, such that information asymmetries and opportunistic behaviors lead to free-riding and shirking (Jensen and Meckling 1976). Beyond the economic losses, agency problems also create costs associated with the need for monitoring, incentive systems, and governance structures (Eisenhardt 1989; Jensen and Meckling 1976). However, agency costs diminish when ownership and management converge, because the principal's and the agent's interests align.

On this basis, many scholars predict that family firms can mitigate opportunistic behaviors and reduce agency and monitoring costs (Jensen and Meckling 1976; Schulze et al. 2002; Le Breton-Miller et al. 2015; Siebels and Knyphausen-Aufseß 2012; Pindado and Requejo 2015). This prediction relies on the expectation of *altruistic* behavior; for example, parents usually act generously to benefit their children. Therefore, family-based altruistic behavior motivates family managers to focus on long-term horizons, promote the family's identity and reputation, and pursue noneconomic goals, without expecting any rewards (Eddleston et al. 2008; Chen and Nowland 2010; Lubatkin et al. 2005; Schulze et al. 2003). Altruistic behavior also aims for the simultaneous enhancement of individual and collective wealth, because it is concentrated in the firm and depends on appropriate management and strategic choices (Schulze et al. 2003; Siebels and Knyphausen-Aufseß 2012; Pindado and Requejo 2015). In turn, several studies have asserted that altruism can be a source of competitive advantage, because it reduces information asymmetries and promotes communication, fostering family commitment and a sense of belonging to the business (Schuleze et al. 2003; Carney 2005; Lubatkin et al. 2007; Eddlestone et al. 2008). An endorsement of this collectivistic view, beyond maintaining family traditions and harmony, also helps prevent the emergence of relationship conflicts, which may be particularly likely in family-managed firms when members belong to different generations or family branches (Eddlestone and Kellermanns 2007; Kellermanns and Eddlestones 2004, 2007).

However, some studies also caution that altruism might tend to transform into *self-control problems* that expose family firms to specific types of agency costs (Schulze et al. 2003; Lubatkin et al. 2007b; Bammens et al. 2011; Siebels and Knyphausen-Aufseß 2012). In this sense, excessively altruistic behavior may allow family priorities to overcome business ones, prompting courses of actions such as nepotism, lavishing excessive perquisites and privileges on employed children, or setting underserved career paths and comparison criteria (Schulze et al. 2003; Chua et al. 2009; Lubatkin et al. 2007a, b). Unlike a traditional moral hazard problem, the self-control challenge (also known as intrapersonal moral hazard) refers to conflicting ideas within a single person, such as an internal struggle by a principal or owner to prioritize family-oriented or entrepreneurial/managerial behavior. In striving for noneconomic objectives and participating in their intrinsic family relationships, family members might lose their self-control and long-term perspective, such that they adopt hazardous actions that threaten firm performance and family wealth (Lubatkin et al. 2005; Bammens et al. 2011; Siebels and Knyphausen-Aufseß 2012). The problem is accentuated in privately held family firms; unlike listed family companies, they are not subject to capital market pressures or active monitoring by shareholders (Anderson and Reeb 2003b; Carney et al. 2013).

Similar to traditional organizations though, family firms might be affected by the principal–principal agency problem (i.e., Type II), which arises between majority and minority shareholders (Villalonga and Amit 2006). The privileged monitoring position of majority owners may expose them to information advantages that they can use to pursue their own interests, to the detriment of other owners. Relative to other types of owners, the family has a stronger potential incentive to expropriate resources, in that when "the large shareholders are an institution such as a bank, an investment fund, or a widely-held corporation, the private benefits of control are diluted among several independent owners" (Villalonga and Amit 2006, p. 2), so their incentive to expropriate resources is minimal. But family firms have great incentives to expropriate resources, because the private benefits of control are concentrated among family members. Type II agency problems thus take different forms in family firms depending on whether minority shareholders belong to the family or not. If they do, the altruism problem resurfaces in second- or later-generation family firms, which often face fragmentation across the separate siblings' family units (Sonfield and Lussier 2004). In sibling partnerships or cousin consortia organizations, excessive altruism by family units, each characterized by its own utility function, leads to an intra-family divergence of interests and favors self-interested actions that disregard overall family (firm) wealth (Schulze et al. 2003; Sonfield and Lussier 2004; Lubatkin et al. 2005; Bammens et al. 2008). Depending on the level of diversity among family members, they might engage in entrenchment or suffer relationship conflicts related to their distinct opinions about strategic issues (e.g., dividend payouts, risk attitude, hiring strategy, incentive systems) (Kellermanns and Eddlestone 2004, 2007; Eddlestone and Kellermanns 2007; Villalonga and Amit 2006).

Substantial research also demonstrates that the entrenchment phenomenon occurs with external minority shareholders too (Claessens et al. 2002; Young et al. 2008; Siebels and Knyphausen-Aufseβ 2012; Carney et al. 2013). With a mixed ownership structure, minority owners who are not part of the dominant family face the so-called expropriation risk. That is, family owners/managers, affected by their excessive altruism, expropriate value from minority shareholders through their behaviors driven by noneconomic goals, which privilege family wealth over firm efficiency or performance (e.g., Villalonga and Amit 2006). Empirical studies verify that, beyond a certain threshold, increasing family managerial ownership enhances the likelihood of managerial entrenchment, increases agency costs, produces less effective governance mechanisms, and hinders performance (e.g., Anderson and Reeb 2003a; Villalonga and Amit 2006; Claessens et al. 2002; Young et al. 2008; Yang 2010; Pindado and Requejo 2015).

This domino effect can be detected and stemmed most easily in publicly listed family firms, in which minority shareholders can discount the expropriated value from the family's equity share (Claessens et al. 2002; Pindado and Requejo 2015). In financial markets, the presence of strong isomorphic norms leads to more severe, effective governance and monitoring systems that limit the degrees of freedom granted to the dominant owning family (La Porta et al. 1998; Carney et al. 2013; Claessens et al. 2002). Conversely, in privately held firms, the pursuit of noneconomic goals and the entrenchment problem are harder to overcome, because no isomorphic forces keep the family from expropriating benefits from non-family minority shareholders. The most troubling scenario occurs when private family firms are characterized by pyramidal ownership structures that favor so-called tunneling activities (Claessens et al. 2002; Siebels and Knyphausen-Aufseß 2012). Johnson et al. (2000, p. 22) use the term *tunneling* to describe the "transfer of resources out of a company to its controlling shareholder," to the detriment of minority owners. For example, private benefits may result from the sale of assets at prices below market value or with terms that are prejudicial to minority owners. In that case, the divesting company's value will decline, because the divested unit is being sold at a price lower than its market value, especially if the sale is made to an acquirer in which the majority owner holds a higher share than in the divesting company.

2.3.2 Resource-Based Theory

Another popular theoretical framework for investigating family businesses is the resource-based view of the firm (RBV). The argument at its core is that different firms reach diverse levels of performance and competitive advantage because they are endowed with different resources (Barney 1991). The value created by different companies depends on how they assemble a bundle of valuable, rare, difficult to imitate and substitute resources (Barney 1991). This resource heterogeneity and complementarity explain differential performance (Barney 1991). Family businesses are complex, multilayered, and multidimensional, such that the RBV has been particularly suitable as a lens of analysis.

Specifically, the idea that a bundle of resources, idiosyncratic to the firm and its environment, produces a company's sustainable competitive advantage is particularly worthwhile for family businesses, because it demands the inclusion of different, idiosyncratic, firm-level characteristics in any analysis (Habberson and Williams 1999). Prior studies thus ascribe the superior performance of family firms to different, relevant traits, such as a family-oriented industrial atmosphere and collective identity (Zellweger et al. 2010), which can foster employee productivity and information exchanges (Ward 1988; Ling and Kellermanns 2010). Furthermore, family ties might increase motivation and commitment to the company's vision and long-term objectives (Chirico 2008; Gomez-Mejia et al. 2007; Zellweger and Astrachan 2008). Other studies cite traits such as community loyalty and respect (Hoffmann et al. 2006), upright reputations (Tagiuri and Davis 1996), and more flexible decision-making processes (Tagiuri and Davis 1996).

Combining these multidimensional effects, some RBV theorists identify *familiness* as the advantage that family firms derive, in terms of their unique and distinctive resources and capabilities that lead to advantage-based rents and high levels of value creation (Habberson and Williams 1999; Chirico et al. 2011a; Chirico and Salvato 2008; Arregle et al. 2007). Extensions of this construct have investigated its components, antecedents, and consequences. For example, by merging the concepts of familiness and organizational social capital (i.e., the relationships between individuals and organizations that enable action and create value; Sharma 2008; Adler and Kwon 2002), scholars propose a social capital model of familiness (Pearson et al. 2008; Sharma 2008) to complement the original, static familiness framework with a critical evolutionary perspective. Familiness is nourished by both the content and flow of social capital. The former is internally oriented; it consists of the networks and relationships that emerge within the organization and the family. The latter is externally oriented and describes networks that develop with actors who interact with the organization and the family but operate primarily outside of the organization (Arregle et al. 2007; Sharma 2008). Content and flow in turn contribute to the emergence of bonding and bridging social capital, respectively. Bonding social capital supports the development of within-family networks and contributes to the formation of a bundle of resources and capabilities. Bridging social capital instead fosters the creation of flows between the family and its environment that produce variation in their social capital and familiness (Sharma 2008; Pearson et al. 2008). If the interaction between content and flow results in a balanced exchange, capital stock will be enriched (distinctive familiness), but if the exchange is unbalanced, capital stock will suffer a decrease (constrictive familiness).

From a similar perspective, an idiosyncratic and unique bundle of resources can originate with the interaction among the family, its members, and the business (Habbershon et al. 2003; Habbershon 2006). Resources and distinctive capabilities get generated by the influence of the external economic and social environment on the family, its members, and the business, as well as by the interactions of each party within the ecosystem (Habbershon 2006). Depending on the interaction process, the idiosyncratic bundle of resources and capabilities can be affected either positively or negatively by family influences.

Offering a further means to explain the evolution of familiness over time and contribute to the integration of the RBV and agency theory, Habbershon (2006) proposes a framework of family-influenced agency interaction. In this framework, younger family businesses (i.e., early stage of the organizational life cycle) benefit from an unbounded familial culture, but older and bigger organizations take advantage of their bounded, well-organized culture, which focuses more on structured monitoring and control mechanisms. In the latter case, the objective is to conserve family value and norms, while avoiding classical agency problems such as adverse selection or moral hazard (Habbershon 2006).

Finally, recent studies investigate the impact of the family's bundle of resources on specific strategic decisions, such as the tendency to engage in franchising (Chirico et al. 2011a) or adopt an entrepreneurial orientation (Kellermanns et al. 2016; Chirico et al. 2011b). Efforts to identify the specific family resources that might foster or deter an entrepreneurial orientation identify factors such as the level of reciprocal altruism (Eddleston et al. 2008) or the amount of parsimony, which might drive family owners to deploy resources with greater care and frugality (Chrisman et al. 2005a).

2.3.3 Stewardship Theory

Stewardship theory focuses on principals and steward-agents (Donaldson and Davis 1991; Davis and Harveston 1999). In contrast with agency theory, which regards the agent as the opportunistic and self-interested party (i.e., economic view), in stewardship theory, the agent is a steward (i.e., humanistic view), characterized by a long-term perspective, commitment, family values, and identification (Eddlestone and Kellermanns 2007; Le Breton-Miller and Miller 2009; Davis et al. 2010; Madison et al. 2016).

In a family business context, a stewardship approach suggests that managers are intrinsically motivated by their identification with the family's business, history, and culture (Le Breton-Miller and Miller 2009; Vallejo-Martos 2009). Family managers feel a sense of belonging and inherently act as stewards, thereby fostering a collectivistic culture (Arregle et al. 2007; Zahra et al. 2008). Non-family managers who are led by steward-family owners also are involved and committed to the firm's prosperity and longevity, laying the foundation for a reciprocal stewardship culture (Vallejo-Martos 2009; Pearson and Maler 2010). This reciprocal stewardship situation is especially prominent in territorially rooted and small enterprises, for which the family, the business, and local wealth are inextricably linked. This collection of binding factors nurtures participative decision making, which then results in the consolidation of governance mechanisms characterized by loyalty and trust (Sirmon and Hitt 2003; Eddleston and Kellermanns 2007; Le Breton-Miller and Miller 2009; Davis et al. 2010). According to Davis et al. (2010), stewardship is the "secret sauce" for creating competitive advantages, derived largely from the influence that pro-organizational and other-serving endeavors have on the family firm's organization (Le Breton-Miller and Miller 2009; Madison et al. 2016).

Therefore, stewardship may appear in the form of three specific expressions that likely occur simultaneously: community, continuity, and connections (Miller et al. 2008). Community refers to the collectivistic culture that encourages commitment, cohesion, loyalty, and senses of belonging and responsibility (Davis et al. 2010; Eddleston and Kellermanns 2007; Miller and Le Breton-Miller 2005; Le Breton-Miller and Miller 2008; Madison et al. 2016). Continuity implies embracing a long-run approach, with the purpose of safeguarding business wealth that corresponds with the resources of the family and the whole organization (Miller and Le Breton-Miller 2005; Le Breton-Miller and Miller 2008). Finally, connections denote relationships with external stakeholders; they are strongly linked to continuity, in that they can help establish long-lasting relationships (Gomez-Meija et al. 2001).

Notwithstanding the supposed inconsistency between agency and stewardship theories, recent efforts have been focused on reconciling these two perspectives in a unique context. In this argument, the applicability of the two approaches may depend on the degree of managers' social embeddedness within the family (Le Breton-Miller and Miller 2009; Le Breton-Miller et al. 2011; Siebels and Knyphausen-Aufseß 2012; Madison et al. 2016).

2.3.4 Behavioral Agency Model

Finally, a pillar of family firm theory relies on the behavioral agency model (BAM) (Tversky and Kahneman 1986; Wiseman and Gomez-Mejia 1998), which predicts that different variables have varying impacts on agents' decision outcomes, and they are not rooted in a rigid or inflexible path (Wiseman and Gomex-Mejia 1998). The only rule that guides decision makers is the preservation of the firm's accumulated, existing endowments (Wiseman and Gomez-Mejia 1998; Gomez-Mejia et al. 2000). Advocates of this view mainly investigate risk-taking behaviors by firm agents, with the argument that a risk-aversion hypothesis actually should be substituted with a loss-aversion one (Wiseman and Gomex-Mejia 1998; Lim et al. 2010; Miller et al. 2014; Le Breton-Miller et al. 2015).

That is, traditional family business literature (e.g., La Porta et al. 1999) posited that wealth concentration in a single firm leads to greater *risk aversion*, such that family firms would be reluctant to pursue potentially high-return investments because of their concentrated ownership position (Morck and Yeung 2003; Gomez-Mejia et al. 2007). On the contrary, according to the BAM, agents modify their risk attitudes depending on their perceptions of prospects for changes to their personal wealth (Wiseman and Gomez-Mejia 1998; Lim et al. 2010). That is, *risk bearing* should relate to the perceived risk imposed on agent wealth (Lim et al. 2010). This reasoning has been widely embraced by scholars who investigate family owners' risky decisions. The widespread assumption that family owners are risk averse accordingly has been replaced with a loss-aversion hypothesis (Lim et al. 2010), which holds that family owners (unlike non-family ones) conceive of loss beyond just economic wealth. The loss they try to avoid extends past simply financial wealth or firm profit to include their socio-emotional endowments (Gomez-Mejia et al. 2007).

A well-established socio-emotional wealth (SEW) model builds on this theory and provides an interesting theoretical formulation for family firm studies (Gomez-Mejia et al. 2007, 2010, 2011; Berrone et al. 2012). Berrone et al. (2012) assert that the inseparable link between family and managerial–business life in family firms represents their primary distinguishing feature: Only in family firms are agents and principals driven by noneconomic goals and affective endowments. As a result, this particular attitude shapes family firm decisions and influences their major strategic choices and policy (Berrone et al. 2012). To preserve their SEW, including social and emotional connections and their resulting benefits, family decision makers thus forgo less compelling, correctly perceived actions, even at the expense of potential higher returns (Gomez-Mejia et al. 2007; Berrone et al. 2012; Naldi et al. 2013; Le Breton-Miller et al. 2015).

Another model proposes that family loss aversion actually encompasses five main dimensions (Berrone et al. 2012):

1. Personal *fulfillment*, derived from successfully running the business, might be threatened by a loss of family control or influence.
2. The *identification* of family members with the firm pushes those members to ensure the preservation of the firm, which often carries their family name. This identification issue has been widely recognized as a distinguishing trait of family firms (Kets de Vries 1993; Gomez-Mejia et al. 2007; Westhead et al. 2001); it constitutes a sort of "overarching construct" that marks the family members who work for the business that sports their family name (Ket de Vries 1993). Through their identification, family members' SEW increases in terms of attachment, perpetuation goals, and long-term perspectives (Westhead et al. 2001; Gomez-Mejia et al. 2007).
3. Social *recognition* stems from family membership, which often allows for the development of strong social ties with employers or external stakeholders (Miller and Le Breton-Miller 2005).

4. Family members also feel an *emotional* attachment, including a sense of belonging, pride, and responsibility toward previous and future generations (Sonfield and Lussier 2004; Kellermanns and Eddleston 2004; Eddleston and Kellermanns 2007).
5. The renewal of family *bonds* to the firm through dynastic succession ultimately explains the family firm's long-term perspective and loss aversion. A family firm cannot be assessed solely using detached economic and financial criteria, because it represents a family history and tradition that should be passed down to later generations (Zellweger and Astrachan 2008). In this sense, the SEW results from ensuring successful career paths in strategic managerial position for siblings and children.

Thus, different families exhibit varying levels of loss aversion, according to their conceptions of wealth. Some might be more concerned about losing reputation, but others worry about losing control (Naldi et al. 2013). The dynamic adjustments to these concerns create a role for *risk bearing* as a critical mediator between how agents frame their wealth prospects and their risk-taking behavior (Lim et al. 2010). Asserting that family agents are risk averse is almost reductive; their risk bearing actually depends on several factors that are hard to disentangle and that affect initiatives in various ways.

2.4   Family Firm Strategic Decisions

Notwithstanding the different theoretical approaches, most empirical studies concur that family owner behaviors, goals, and interests influence the firm's strategic decisions and thus its performance (Pindado and Requejo 2015). Previous studies often focus on specific strategic decisions, such as diversification (Gomez-Mejia et al. 2007, 2010; Anderson and Reeb 2003a; Schmid et al. 2015), internationalization (Fernández and Nieto 2006; Gedajlovic et al. 2004), financing strategies (Anderson and Reeb 2003b; Pindado et al. 2012; Mishra and McConaughy 1999), investment policies (Pindado et al. 2011), R&D investments, or innovation management (De Massis et al. 2015;

De Massis et al. 2013; Carnes and Ireland 2013; Matzler et al. 2015; Chrisman et al. 2015). But as a common basis, recent studies note that family firms are consistently affected by the "mixed gamble dilemma" (Gomez-Mejia et al. 2015). That is, when they must make a strategic decision, family firms exhibit their risk-bearing tendencies. Because family owners tend to be more concerned with loss aversion, rather than risk aversion, they have difficulty resolving the trade-off between their financial and SEW considerations (Chrisman and Paterl 2012; Gomez-Mejia et al. 2011). The family members might focus more on preventing the firm's vulnerability, even if the resulting actions lead to "below-target" performance (Gomez-Mejia et al. 2007, 2015). In other words, they are more willing to undertake *venturing* risks than *performance hazard* risks, because they seek to preserve the status quo and their socio-emotional endowments (Gomez-Mejia et al. 2007, 2015). In turn, this loss aversion attitude affects various strategic decisions, as detailed in the following sections, while Chap. 3 provides a comprehensive and integrative review of existing research on the management of technological innovation by family firms.

2.4.1 Diversification

To address their financial considerations, family firms should undertake diversification strategies to spread their business portfolio and risk (Gomez-Mejia et al. 2010). Furthermore, next-generation prosperity, in SEW terms, is closely related to the persistence of the firm, so reducing risk should be a primary purpose for family members (Casson 1999). Despite these rationales suggesting that family firms should undertake diversification strategies, several studies highlight opposite findings. Anderson and Reeb (2003b) find that family ownership relates significantly to diversification choices—defined as a decision to combine business units from separate industries under a single firm's roof (Schmid et al. 2015)—and that family firms engage in significantly (15%) less corporate diversification. Consistent with stewardship theory, they attribute this finding to family managers' commitment to ensure the firm's competitive advantage, such that they "avoid diversification because of its substantial negative effects" (Anderson and Reeb 2003b, p. 659). Yet according to the BAM, the reduced level of diversification instead results from the loss of SEW, regardless of the negative performance implications (Berrone et al. 2012; Gomez-Mejia et al. 2007, 2010). That is, diversification requires funding, which is unlikely to come from the family. Thus, it requires access to external resources, whether through the entrance of external shareholders or debt capital. Both options imply some loss of authority, control, and influence (Schulze et al. 2003; Schmid et al. 2015), so these diversification activities represent a hazard to SEW, in terms of family control and power, by introducing new players into the organizational routine and challenging well-established practices (Gomez-Mejia et al. 2010). From this standpoint, family management is not a strength but rather a relative weakness in terms of financial performance, because managers avoid financially advantageous operations, just to preserve their control. Mishra and McCounaghy (1999) concur that family-owned firms prefer to grow by leveraging their internal resources, even at the expense of profitable opportunities, rather than increasing their external dependence (Casson 1999).

In addition to opening the family to external influences in terms of ownership and governance structure, diversifying into new industries also increases the need for non-family, professional management that possesses the competences needed to succeed in the new business (Arregle et al. 2012). According to the RBV, as the number of external executives rises, the level of familiness—from which family firms draw their competitive advantage—decreases (Habbershon and Williams 1999; Morck and Yeung 2003; Morck et al. 2000). The detachment from family values and norms, which were instituted by the founder and rooted in the family and its history, thus may exert a negative impact on firm performance (Habbershon and Williams 1999). On the flipside, avoiding diversification and failing to introduce external managers could have detrimental effects too, because it threatens the organization with inertia (Salvato and Melin 2008; Cannella et al. 2008; Zellweger et al. 2010; Chirico and Nordqvist 2010). When a firm is trapped by its rigidities, standard policies, and routines, it cannot respond to changing environments or seize profitable opportunities

(Schulze et al. 2003). From this perspective, organizational inertia obstructs the complex diversification process, which implies a change not only to the industries covered but also to the business processes and *modus operandi* (Eisenmann 2002; Gomez Mejia et al. 2010; Binacci et al. 2016). With this reasoning, family firms that refuse to hire external managers, with the objective of conserving their "dynasty," may exacerbate problems associated with nepotism, free riding, and adverse selection (Barnett and Kellermans 2006). They prefer to limit the top management team to family members, regardless of their professional skills or competences, even though widening the circle to external talented managers could lead to firm growth and greater wealth (Gomez-Mejia et al. 2010; Cannella et al. 2008; Zellweger et al. 2010).

2.4.2 Internationalization

The degree to which a firm internationalizes its reference markets or operations is another key decision that family firms face. As a type of diversification, expansion to internationally diverse markets can help mitigate the level of risk borne by an organization (Kim et al. 1993; Hitt et al. 1997; Sanders and Carpenter 1998), because it gets spread over different countries, which lowers the overall level of total and systematic risk (Kogut 1985). Internationalization also can benefit firm performance, because it allows the firm to move beyond the boundaries of its domestic market, exploit demand in various countries, avoid some tariffs, and leverage its core competences in different marketplaces (Sanders and Carpenter 1998).

However, the distinguishing features of family firms have prompted varied predictions about the outcomes of their internationalization, depending on the theories used to explain the results (Gomez-Mejia et al. 2010). Similar to diversification, family firms might renounce profitable internationalization opportunities to avoid the loss of SEW (Gomez-Mejia et al. 2010; Pindado and Requejo 2015). Here again, reduced internationalization activity might reflect the owners' desire to concentrate ownership and managerial control within the family (Gomez-Mejia et al. 2010). The intrinsic complexity and massive financial resources needed for any internationalization effort suggest that such operations require input from external investors, representing a potential threat to family power.

Internationalization in culturally distant countries in particular may prevent family managers from exploiting their best practices and organizational routines in new markets, which would require them to hire external non-family managers with distinct competencies (Hofstede 1980; Hitt et al. 1997). Given that family firms prefer to keep their top management team closed, they likely avoid complex international activities (Gedajlovic et al. 2004). In a related sense, internationalization demands effort to build external ties with new customers and other critical stakeholders in the new market (e.g., suppliers, institutions, credit systems). Detaching from a local territory is often one of the greatest threats facing family firms, because they derive most of their social status, identification, and recognition—that is, their SEW endowment (Gomez-Mejia et al. 2010)—from the local environment in which the firm was founded. In turn, several empirical studies affirm that family control and ownership are negatively associated with internationalization strategies (Gomez-Mejia et al. 2010; Eberhard and Craig 2013; Pindado and Requejo 2015).

2.4.3 Financing and Investment Strategies

Because of their goal to preserve control, family firms often have less levered capital structures and make little use of debt capital, to avoid handing over more control than needed to credit providers (Mishra and McConaughy 1999). The basis for this family debt aversion resides not only in the fear of a loss of control but also in the increased likelihood of family conflict associated with debt acquisition (Gomez-Mejia et al. 2010, 2011; Carney et al. 2013). Despite substantial research agreement about this behavioral attitude, no shared opinion exists regarding its effects on firm profitability or performance. In other words, the debate about whether debt aversion represents a competitive advantage or a disadvantage for family firms remains open (Carney et al. 2013). At the basis of this dispute are two main theoretical approaches.

First, the BAM predicts that family capital structure decisions are driven by noneconomic goals, such as loss of SEW, so they avoid riskier but potentially profitable growth opportunities, which ultimately harms family firm performance (Mishra and McConaughy 1999; Chandler 1990). Second, stewardship theory highlights the long-term perspective adopted by family managers, which pushes them to avoid debt financing because it might increase the risk of bankruptcy and endanger their long-term profitability (Arregle et al. 2007; Miller et al. 2008).

Another pertinent decision refers to investment policies. Family firms generally have greater investment and cash flow sensitivity, such that they often prefer more slow and organic growth, rather than rapid, financially challenging acquisitions of new business (Pindado and Requejo 2015; Gomez-Mejia et al. 2015). Acquisitions, in addition to representing challenges to cash flows, also imply an openness to entrepreneurial notions and cultures (Zellweger et al. 2012) and granting more control to external stakeholders with unique entrepreneurial competencies (Gomez-Mejia et al. 2010, 2015). Finally, they threaten a loss of firm reputation, due to the dynamism that characterizes mergers of products/services, routines, and resources, all of which tend to create confusion and a lack of identity (Deephouse and Jaskiewicz 2013; Gomez-Mejia et al. 2015).

References

Adler, P. S., & Kwon, S. (2002). Social capital: Prospects for a new concept. *Academy of Management Review, 27*(1), 17–40.

Anderson, R. C., & Reeb, D. M. (2003a). Founding-family ownership and firm performance: Evidence from the S&P 500. *Journal of Finance, 58*(3), 1301–1328.

Anderson, R. C., & Reeb, D. M. (2003b). Founding-family ownership, corporate diversification and firm leverage. *Journal of Law and Economics, 46*, 653–680.

Arregle, J., Hitt, M. A., Sirmon, D. G., & Very, P. (2007). The development of organizational social capital: Attributes of family firms. *Journal of Management Studies, 44*(1), 73–95.

Arregle, J. L., Naldi, L., Nordqvist, M., & Hitt, M. A. (2012). Internationalization of family-controlled firms: A study of the effects of external involvement in governance. *Entrepreneurship Theory and Practice, 36*(6), 1115–1143.

Astrachan, J. H., & Shanker, M. C. (2003). Family businesses' contribution to the U.S. economy: A closer look. *Family Business Review, 16*(3), 211–219.

Astrachan, J. H., Klein, S. B., & Smyrnios, K. X. (2002). The F-PEC scale of family influence: A proposal for solving the family business definition problem. *Family Business Review, 15*(1), 45–58.

Bammens, Y., Voorderckers, W., & Van Gils, A. (2008). Board of directors in family firms: A generational perspective. *Small Business Economics, 31,* 163–180.

Bammens, Y., Voordeckers, W., & Van Gils, A. (2011). Boards of directors in family businesses: A literature review and research agenda. *International Journal of Management Reviews, 13,* 134–152.

Barnett, T., & Kellermanns, F. W. (2006). Are we family and are we treated as family? Nonfamily employees' perceptions of justice in the family firm. *Entrepreneurship Theory & Practice, 3,* 837–854.

Barney, J. (1991). Firm resources and sustained competitive advantage. *Journal of Management, 27*(1), 99–120.

Berrone, P., Cruz, C., & Gomez-Mejia, L. R. (2012). Socioemotional wealth in family firms. Theoretical dimensions, assessment approaches and agenda for future research. *Family Business Review, 25*(3), 258–279.

Binacci, M., Peruffo, E., Oriani, R., & Minichilli, A. (2016). Are all non-family managers (NFMs) equal? The impact of NFM characteristics and diversity on family firm performance. *Corporate Governance: An International Review, 24,* 569–583.

Botero, I. C., Cruz, C., De Massis, A., & Nordqvist, M. (2015). Family business research in the European context. *European Journal of International Management, 9*(2), 139–159.

Cannella, A. A., Park, J., & Lee, H. (2008). Top management team functional background diversity and firm performance: Examining the roles of team member collocation and environmental uncertainty. *Academy of Management Journal, 51*(4), 768–784.

Carnes, C. M., & Ireland, R. D. (2013). Familiness and innovation: Resource bundling as the missing link. *Entrepreneurship Theory & Practice, 37,* 1399–1419.

Carney, M. (2005). Corporate governance and competitive advantage in family-controlled firms. *Entrepreneurship Theory & Practice, 29,* 249–265.

Carney, M., Van Essen, M., Gedajlovic, E. R., & Heugens, P. P. M. A. R. (2013). What do we know about private family firms? A meta-analytical review. *Entrepreneurship Theory and Practice, 39*(3), 513–544.

Carney, M., Van Essen, M., Gedajlovic, E. R., & Heugens, P. (2015). What do we know about private family firms? A meta-analytical review. *Entrepreneurship Theory & Practice, 39*(3), 513–544.

Casson, M. (1999). The economics of the family firm. *Scandinavian Economic History Review, 47*(1), 10–23.

Chandler, A. D. (1990). *Scale and scope: The dynamics of industrial competition.* Cambridge, MA: Harvard University Press.

Chen, E. T., & Nowland, J. (2010). Optimal board monitoring in family-owned companies: Evidence from Asia. *Corporate Governance: An International Review, 18,* 3–17.

Chirico, F. (2008). Knowledge accumulation in family firms. Evidence from four case studies. *International Small Business Journal, 26,* 433–462.

Chirico, F., & Nordqvist, M. (2010). Dynamic capabilities and transgenerational value creation in family firms: The role of organizational culture. *International Small Business Journal, 28*(5), 487–504.

Chirico, F., & Salvato, C. (2008). Knowledge integration and dynamic organizational adaptation in family firms. *Family Business Review, 21,* 169–181.

Chirico, F., Ireland, R. D., & Sirmon, D. G. (2011a). Franchising and the family firm: Creating unique sources of advantage through 'familiness'. *Entrepreneurship Theory and Practice, 35*(3), 483–501.

Chirico, F., Sirmon, D. G., Sciascia, S., & Mazzolla, P. (2011b). Resource orchestration in family firms: Investigating how entrepreneurial orientation, generational involvement and participative strategy affect performance. *Strategic Entrepreneurship Journal, 5*(307), 326.

Chrisman, J. J., & Patel, P. J. (2012). Variations in R&D investments of family and non-family firms: Behavioral agency and myopic loss aversion perspectives. *Academy of Management Journal, 55,* 976–997.

Chrisman, J. J., Chua, J. H., & Litz, R. (2003). A unified systems perspective of family firm performance: An extension and integration. *Journal of Business Venturing, 18,* 467–472.

Chrisman, J. J., Chua, J. H., & Steier, L. (2005a). Sources and consequences of distinctive familiness: An introduction. *Entrepreneurship Theory & Practice, 29,* 237–247.

Chrisman, J. J., Chua, J. H., & Sharma, P. (2005b). Trends and directions in the development of strategic management theory of the family firm. *Entrepreneurship Theory & Practice, 29,* 555–575.

Chrisman, J. J., Chua, J. H., Pearson, A. W., & Barnett, T. (2012). Family involvement, family influence, and family-centered non-economic goals in small firms. *Entrepreneurship Theory & Practice, 36*(2), 267–293.

Chrisman, J. J., Chua, J. H., De Massis, A., Frattini, F., & Wright, M. (2015). The ability and willingness paradox in family firm innovation. *Journal of Product Innovation Management, 32,* 310–318.

Chua, J. H., Chrisman, J. J., & Sharma, P. (1999). Defining the family business by behavior. *Entrepreneurship Theory & Practice, 23*(4), 19–39.

Chua, J. H., Chrisman, J. J., & Bergiel, E. B. (2009). An agency theoretic analysis of the professionalized family firm. *Entrepreneurship Theory & Practice, 33,* 355–372.

Chua, J. H., Chrisman, J. J., Steier, L., & Rau, S. (2012). Sources of heterogeneity in family firms—An introduction. *Entrepreneurship Theory and Practice, 36,* 1103–1113.

Claessens, S., & Tzioumis, K. (2006). Ownership and financing structures of listed and large non-listed corporations. *Corporate Governance: An International Review, 14,* 266–276.

Claessens, S., Djankov, S., Fan, J. P. H., & Lang, L. H. P. (2002). Disentangling the incentive and entrenchment effects of large shareholdings. *Journal of Finance, 57,* 2741–2771.

Davis, J., & Harveston, P. (1999). In the founder's shadow: Conflict in the family firm. *Family Business Review, 12*(1), 311–323.

Davis, J. H., Allen, M. R., & Hayes, H. D. (2010). Is blood thicker than water? A study of stewardship perceptions in family business. *Entrepreneurship Theory and Practice, 34,* 1093–1116.

De Massis, A., Frattini, F., & Lichtenthaler, U. (2013). Research on technological innovation in family firms: Present debates and future directions. *Family Business Review, 26,* 10–31.

De Massis, A., Di Minin, A., & Frattini, F. (2015). Family-driven innovation. *California Management Review, 58*(1), 5–19.

Deephouse, D. L., & Jaskiewicz, P. (2013). Do family firms have better reputations than non-family firms? An integration of socioemotional wealth and social identity theories. *Journal of Management Studies, 50,* 337–360.

Donaldson, L., & Davis, J.H., (1991). Stewardship Theory or Agency Theory: CEO Governance and Shareholder Returns. *Australian Journal of Management, 16*, 49–64.

Eberhard, M., & Craig, J. (2013). The evolving role of organisational and personal networks in international market venturing. *Journal of World Business, 48*, 385–397.

Eddleston, K., & Kellermanns, F. (2007). Destructive and productive family relationships: A stewardship theory perspective. *Journal of Business Venturing, 22*(4), 545–565.

Eddleston, K. A., Kellermanns, F. W., & Sarathy, R. (2008). Resource configuration in family firms: Linking resources, strategic planning and technological opportunities to performance. *Journal of Management Studies, 45*, 26–50.

Eisenhardt, M. K. (1989). Agency theory: An assessment and review. *Academy of Management Review, 14*(1), 57–74.

Eisenmann, T. R. (2002). The effects of CEO equity ownership and firm diversification on risk taking. *Strategic Management Journal, 23*, 513–534.

Fernández, Z., & Nieto, M. J., (2006). Impact of ownership on the international involvement of SMEs. *Journal of International Business Studies, 37*, 340–351.

Fiegener, M. K. (2010). Locus of Ownership and Family Involvement in Small Private Firms. *Journal of Management Studies, 47*(2), 296–321.

Gedajlovic, E., Lubatkin, M. H., & Schulze, W. S. (2004). Crossing the threshold from founder management to professional management: A governance perspective. *Journal of Management Studies, 41*(5), 899–912.

Global Family Business Index, University of St. Gallen, Center for Family Business, EY Family Business Yearbook. (2016). Available at:http://familybusinessindex.com/.

Gomez-Mejia, L. R., Welbourne, T. R., & Wiseman, R. (2000). The role of risk sharing and risk taking under gainsharing. *Academy of Management Review, 23*, 492–509.

Gomez-Mejia, L. R., Nuñez-Nickel, M., & Gutierrez, I. (2001). The role of family ties in agency contracts. *Academy of Management Journal, 44*, 81–95.

Gomez-Mejia, L. R., Haynes, K., Nuñez-Nickel, M., Jacobson, K. J. L., & Moyano-Fuentes, J. (2007). Socioemotional wealth and business risks in family-controlled firms: Evidence from Spanish olive oil mills. *Administrative Science Quarterly, 52*(1), 106–137.

Gomez-Mejia, L. R., Makri, M., & Larraza Kintana, M. (2010). Diversification decisions in family controlled firms. *Journal of Management Studies, 47*(2), 223–252.

Gomez-Mejia, L. R., Cruz, C., Berrone, P., & De Castro, J. (2011). The bind that ties: Socioemotional wealth preservation in family firms. *Academy of Management Annals, 5*(1), 653–707.

Gomez-Mejia, L. R., Patel, P. C., & Zellweger, T. M. (2015). In the horns of the dilemma: Socioemotional wealth, financial wealth, and acquisitions in family firms. *Journal of Management*. doi:10.1177/0149206315614375.

Habbershon, T. G. (2006). Commentary: A framework for managing the familiness and agency advantages in family firms. *Entrepreneurship Theory & Practice, 30*(6), 879–886.

Habbershon, T. G., Williams, M. L., & MacMillan, I. C. (2003). A unified systems perspective of family firm performance. *Journal of Business Venturing, 18,* 451–465.

Habberson, T. G., & Williams, M. L. (1999). Are source-based framework for assessing the strategic advantages of family firms. *Family Business Review, 12*(1), 1–25.

Hitt, M. A., Hoskisson, R. E., & Kim, H. (1997). International diversification: Effects on innovation and firm performance in product-diversified firms. *Academy of Management Journal, 40*(4), 767–798.

Hoffman, J. H., Hoelscher, M., & Sorenson, R. (2006). Achieving sustained competitive advantage: A family capital theory. *Family Business Review, 19,* 135–145.

Hofstede, G. (1980). *Culture's consequences: International differences in work-related values.* Beverly Hills, CA: Sage.

Jensen, M. C., & Meckling, W. H. (1976). Theory of the firm: Managerial behavior, agency costs and ownership structure. *Journal of Financial Economics, 3*(4), 305–360.

Johnson, S., La Porta, R., Lopez-De-Silanes, F., & Shleifer, A. (2000). Tunneling. *American Economic Association, 90,* 22–27.

Kellermanns, F. W., & Eddleston, K. (2004). Feuding families: When conflict does a family firm good. *Entrepreneurship Theory & Practice, 28*(3), 209–228.

Kellermanns, F. W., & Eddleston, K. A. (2007). A family perspective on when conflict benefits family firm performance. *Journal of Business Research, 60*(10), 1048–1057.

Kellermanns, F. W., Walter, J., Crook, T. T., Kemmerer, B., & Narayanan, V. (2016). The resource-based view in entrepreneurship: A content-analytical comparison of researchers' and entrepreneurs' views. *Journal of Small Business Management, 54*(1), 26–48.

Kets de Vries, M. F. R. (1993). The dynamics of family controlled firms: The good and the bad news. *Organizational Dynamics, 21,* 59–71.

Kim, C. W., Hwang, P., & Burgers, W. P. (1993). Multinationals' diversification and the risk-return trade-off. *Strategic Management Journal, 14*(4), 275–286.

Klein, S. B. (2000). Family businesses in Germany: Significance and structure. *Family Business Review, 13,* 157–182.

Kogut, B. (1985). Designing global strategies: Profiting from operating flexibility. *Sloan Management Review, 27*(1), 27–38.

La Porta, R., Lopez-de-Silanes, F., Shleifer, A., & Vishny, R. W. (1998). Law and finance. *Journal of Political Economy, 106*(6), 1113–1155.

La Porta, R., Lopez-De-Silanes, F., & Shleifer, A. (1999). Corporate ownership around the world. *Journal of Finance, 54*(2), 471–516.

Lansberg, I. (1988). The succession conspiracy. *Family Business Review, 1*(2), 119–142.

Le Breton-Miller, I., & Miller, D. (2008). To grow or to harvest? Governance, strategy and performance in family and lone founder firms. *Journal of Strategy and Management, 1*(1), 41–56.

Le Breton-Miller, I., & Miller, D. (2009). Agency vs. stewardship in public family firms: A social embeddedness reconciliation. *Entrepreneurship Theory & Practice, 33,* 1169–1191.

Le Breton-Miller, I., Miller, D., & Lester, R. H. (2011). Stewardship or agency? A social embeddedness reconciliation of conduct and performance in public family businesses. *Organization Science, 22,* 704–721.

Le Breton-Miller, I., Miller, D., & Bares, F. (2015). Governance and entrepreneurship in family firms: Agency, behavioral agency and resource-based comparisons. *Journal of Family Business Strategy, 6*(1), 58–62.

Lichtenthaler, U., & Muethel, M. (2012). The impact of family involvement on dynamic innovation capabilities: Evidence from German manufacturing firms. *Entrepreneurship Theory and Practice, 36*(6), 1235–1253.

Lim, E. N. K., Lubatkin, M. H., & Wiseman, R. M. (2010). A family firm variant of the behavioral agency theory. *Strategic Entrepreneurship Journal, 4,* 197–211.

Ling, Y., & Kellermanns, F. W. (2010). The effects of family firm specific diversity: The moderating role of information exchange frequency. *Journal of Management Studies, 47*, 332–344.

Lubatkin, M. H., Schulze, W. S., Ling, Y., & Dino, R. N. (2005). The effects of parental altruism on the governance of family-managed firms. *Journal of Organizational Behavior, 26*, 313–330.

Lubatkin, M. H., Ling, Y., & Schulze, W. S. (2007a). An organizational justice-based view of self-control and agency costs in family firms. *Journal of Management Studies, 44*(6), 955–971.

Lubatkin, M. H., Durand, R., & Ling, Y. (2007b). The missing lens in family firm governance theory: A self-other typology of parental altruism. *Journal of Business Research, 60*(10), 1022–1029.

Madison, K., Holt, D. T., Kellermanns, F. W., & Ranft, A. L. (2016). Viewing family firm behavior and governance through the lens of agency and stewardship theories. *Family Business Review, 29*(1), 65–93.

Matzler, K., Veider, V., Hautz, J., & Stadler, C. (2015). The impact of family ownership, management, and governance on innovation. *Journal of Product Innovation Management, 32*, 319–333.

Melin, L., & Nordqvist, M. (2007). The reflexive dynamics of institutionalization: The case of the family business. *Strategic Organization, 5*(3), 321–333.

Miller, D., & Le Breton-Miller, I. (2005). *Managing for the long run*. Boston, MA: Harvard Business School Press.

Miller, D., Le Breton-Miller, I., & Scholnick, B. (2008). Stewardship vs. stagnation: An empirical comparison of small family and non-family businesses. *Journal of Management Studies, 45*, 51–78.

Miller, D., Le Breton-Miller, I., Minichilli, A., Corbetta, G., & Pittino, D. (2014). When do non-family CEOs outperform in family firms? Agency and behavioural agency perspectives. *Journal of Management Studies, 51*(4), 547–572.

Minichilli, A., Corbetta, G., & MacMillan, I. (2010). Top management team in family-controlled companies: "Familiness","faultline", and their impact on financial performance. *Journal of Management Studies, 47*(2), 205–222.

Mishra, C. S., & McConaughy, D. L. (1999). Founding family control and capital structure: The risk of loss of control and the aversion to debt. *Entrepreneurship Theory and Practice, 23*(4), 53–64.

Morck, R., & Yeung, B. (2003). Agency problems in large family business groups. *Entrepreneurship Theory and Practice, 27*(4), 367–382.

Morck, R., & Yeung, B. (2004). Family control and the rent-seeking society. *Entrepreneurship Theory & Practice, 28,* 391–409.

Morck, R., Stangeland, D., & Yeung, B. (2000). Inherited wealth, corporate control, and economic growth: The Canadian disease. In R. Morck (Ed.), *Concentrated corporate ownership* (pp. 319–369). Chicago: University of Chicago Press.

Naldi, L., Cennamo, C., Corbetta, G., & Gomez-Mejia, L. (2013). Preserving socioemotional wealth in family firms: Asset or liability? The moderating role of business context. *Entrepreneurship Theory and Practice, 37*(6), 1341–1360.

Nordqvist, M., Sharma, P., & Chirico, F. (2014). Family firm heterogeneity and governance: A configuration approach. *Journal of Small Business Management, 52*(2), 192–209.

Pearson, A. W., & Marler, L. E. (2010). A leadership perspective of reciprocal stewardship in family firms. *Entrepreneurship Theory and Practice, 34*(6), 1117–1124.

Pearson, A., Carr, J. C., & Shaw, J. C. (2008). Toward a theory of familiness: A social capital perspective. *Entrepreneurship Theory and Practice, 32*(6), 949–969.

Pindado, J., & Requjo, I. (2015). Family business performance from a governance perspective: A review of empirical research. *International Journal of Management Reviews, 17,* 279–311.

Pindado, J., Requejo, I., & de la Torre, C. (2011). Family control and investment—Cash flow sensitivity: Empirical evidence from the Euro zone. *Journal of Corporate Finance, 17,* 1389–1409.

Pindado, J., Requejo, I., & de la Torre, C. (2012). Do family firms use dividend policy as a governance mechanism? Evidence from the Euro zone. *Corporate Governance: An International Review, 20,* 413–431.

Prencipe, A., Bar-Yosef, S., Mazzola, P., & Pozza, L. (2011). Income smoothing in family-controlled companies: Evidence from Italy. *Corporate Governance: An International Review, 19,* 529–546.

Salvato, C., & Melin, L. (2008). Creating value across generations in family-controlled businesses: The role of family social capital. *Family Business Review, 21*(3), 259–276.

Sanders, W. M. G., & Carpenter, M. A. (1998). Internationalization and firm governance: The roles of CEO compensation, top team composition, and board structure. *Academy of Management Journal, 41,* 158–178.

Schmid, T., Ampenberger, M., Kaserer, C., & Achleitner, A. K. (2015). Family firm heterogeneity and corporate policy: Evidence from diversification decisions. *Corporate Governance: An International Review, 23*(3), 285–302.

Schulze, W. S., Lubatkin, M. H., & Dino, R. N. (2002). Altruism, agency and competitiveness in family firms. *Managerial and Decision Economics, 23,* 247–259.

Schulze, W. S., Lubatkin, M. H., & Dino, R. N. (2003). Exploring the agency consequences of ownership dispersion among the directors of private family firms. *Academy of Management Journal, 46,* 179–194.

Shanker, M. C., & Astrachan, J. H. (1996). Myths and realities: Family businesses' contribution to the US economy—A framework for assessing family business statistics. *Family Business Review, 9*(2), 107–123.

Sharma, P. (2004). An overview of the field of family business studies: Current status and directions for the future. *Family Business Review, 17,* 1–36.

Sharma, P. (2008). Familiness: Capital stocks and flows between family and business. *Entrepreneurship Theory & Practice, 32,* 971–977.

Siebels, J.-F., & zu Knyphausen-Aufseß, D. (2012). A review of theory in family business research: The implications for corporate governance. *International Journal of Management Reviews, 14,* 280–304.

Sirmon, D. G., & Hitt, M. A. (2003). Managing resources: Linking unique resources, management, and wealth creation in family firms. *Entrepreneurship Theory and Practice, 27,* 339–358.

Sirmon, D. G., Arregle, J. L., Hitt, M. A., & Webb, J. W. (2008). The role of family influence in firms' strategic responses to threat of imitation. *Entrepreneurship Theory & Practice, 32,* 979–998.

Sonfield, M. C., & Lussier, R. N. (2004). First-, Second-, and Third-Generation family firms: A comparison. *Family Business Review, 17*(3), 189–201.

Tagiuri, R., & Davis, J. A. (1996). Bivalent attributes of the family firm. *Family Business Review, 9*(2), 199–208.

The Economist. (2015). To have and to hold. *The Economist, Special Report,* 1–14.

Tversky, A., & Kahneman, D. (1986). Rational choice and the framing of decisions. *Journal of Business, 59*(4), 251–278.

Vallejo-Martos, M. C. (2009). The effects of commitment of non-family employees of family firms from the perspective of stewardship theory. *Journal of Business Ethics, 87*(3), 379–390.

Verbeke, A., & Kano, L. (2012). The transaction cost economics theory of the family firm: Family-based human asset specificity and the bifurcation bias. *Entrepreneurship Theory & Practice, 36,* 1183–1205.

Villalonga, B., & Amit, R. (2006). How do family ownership, control and management affect firm value? *Journal of Financial Economics, 80*(2), 385–417.

Ward, J. L. (1988). The special role of strategic planning for family businesses. *Family Business Review, 1*(2), 105–117.

Westhead, P., & Howorth, C. (2006). Ownership and management issues associated with family firm performance and company objectives. *Family Business Review, 19,* 301–316.

Westhead, P., Cowling, M., & Howorth, C. (2001). The development of family companies: Management and ownership issues. *Family Business Review, 14*(4), 369–385.

Wiseman, R. M., & Gomez-Mejia, L. R. (1998). A behavioral agency model of managerial risk taking. *Academy of Management Journal, 23*(1), 133–153.

Yang, M.-L. (2010). The impact of controlling families and family CEOs on earnings management. *Family Business Review, 23,* 266–279.

Young, M. N., Peng, M. W., Ahlstrom, D., Bruton, G. D., & Jiang, Y. (2008). Corporate governance in emerging economies: A review of the principal–principal perspective. *Journal of Management Studies, 45,* 196–220.

Zahra, S. A., Hayton, J. C, Neubaum, D. O., Dibrell, C., & Craig, J. (2008). Culture of family commitment and strategic flexibility: The moderating effect of stewardship. *Entrepreneurship: Theory and Practice, 32*(6), 1035–1054.

Zellweger, T. M., & Astrachan, J. H. (2008). On the emotional value of owning a firm. *Family Business Review, 21,* 347–363.

Zellweger, T., Eddleston, K. H., & Kellermanns, F. W. (2010). Exploring the concept of familiness: Introducing family firm identity. *Journal of Family Business Strategy, 1*(1), 54–63.

Zellweger, T. M., Nason, R. S., & Nordqvist, M. (2012). From Longevity of Firms to Transgenerational Entrepreneurship of Families. *Family Business Review, 25*(2), 136–155.

3

Innovation in Family Firms: Critical Review of Theoretical and Empirical Literature

Abstract This chapter complements the analysis of the idiosyncratic features of family firms developed in Chap. 2 by leveraging innovation literature, in an effort to provide a comprehensive, integrative review of existing research on the topic. This chapter introduces the role and characteristics of technological innovation within contemporary firms, as well as the links among technological innovation, value creation, competitive advantage, and firm survival. Starting from this framework, it maps the most critical decision-making elements associated with the management of technological innovation by family firms. The comparison of several, often divergent viewpoints from different streams of literature produces a systematic account of theories and empirical findings pertaining to the management of innovation in family firms.

Keywords Family firms · Technological innovation · R&D · Innovation performance

© The Author(s) 2017
A. Perri, *Family Business and Technological Innovation*,
DOI 10.1007/978-3-319-61596-7_3

3.1 Technological Innovation in Contemporary Organizations

Innovation is pivotal to the growth and survival of all contemporary organizations. Since pioneering work by Schumpeter (1934), scholars, practitioners, and policy makers have acknowledged that an effective innovation strategy, combined with an appropriate bundle of tangible and intangible resources and an organization that has been designed to ensure its effective implementation, is one of the most reliable pathways to value creation. The widespread interest in innovation in the field of management stems from its capacity to enable the firm to be the first to bring a given offer to the market, which is a critical source of strategic advantage (Lieberman and Montgomery 1988). This principle is truly vital today, in that both firm-internal and firm-external environments tend to be turbulent and uncertain, requiring firms to exhibit strong capacities to modify and upgrade their products and processes. A key driver of differential firm performance thus is the ability to introduce innovations that produce economic rents (e.g., Lawless and Anderson 1996; Christensen 1997).

Technological innovation is the process through which an entrepreneur combines various inputs to exploit opportunities to commercialize an output (e.g., product, service, process, or business model; Drucker 1985). The process involves many decisions, as detailed effectively in the vast literature on technology and innovation management. In particular, this field has closely investigated the choice to invest in R&D, an inevitable input to technological innovation. More generally, scholars address how companies can assimilate and organize a range of diverse technological inputs to create original inventions or facilitate successful new product development. Research also indicates when and how firms might choose to rely on external technology sources. The investigations of these and other factors seek to identify the key drivers of innovation performance. Finally, scholars have tried to understand how companies can manage imitation threats and appropriate value from their technological innovation.

Like any company, family firms can benefit from innovation. Yet the specific characteristics of these organizations lead to a significant degree of heterogeneity in their innovative behaviors, compared with those of non-family firms. Corporate governance, which reflects the combination of mechanisms that determine how a firm's resources will be allocated and its returns distributed, helps explain some of these variations in a firm's innovation behavior. However, only in recent decades have corporate governance institutions and mechanisms started to be investigated as influences on a firm's technological decisions and innovation performance (Bushee 1998; Coriat and Weinstein 2002; Hall and Soskice 2001; Lee and O'Neil 2003; O'Sullivan 2000; Tylecote and Ramirez 2006; Munari et al. 2010; Choi et al. 2012). This chapter, therefore, offers a comprehensive review of literature in this realm, with a focus on family governance, along with a systematic summary of the most important findings and theoretical insights that may inspire further developments in this research area.

3.2   Literature Review Methodology

To identify studies to include in the literature review, the starting point was an article search of the ISI Web of Science database. This database offers well-accepted reliability in terms of its selection standards and is widely diffused among the academic community (Klang et al. 2014; Perri and Peruffo 2016). The first step entailed a search for two keywords, "family" and "innovation," in the full text of any academic articles classified in the subject fields of "Management" and "Business." A close examination of the title, abstract, and when necessary, full text of each of the 288 articles thus identified eliminated any texts that were not relevant to the topic, whether because the study did not focus on innovation in family firms or because it lacked clarity in its definition of the scope and meaning of the phenomenon it was analyzing (e.g., Strutzenberger and Ambos 2014). This procedure resulted in a set of 71 articles, which further revealed 11 articles that were cited frequently as prior art but that did not emerge out of the search process. Adding these pertinent studies created a sample of 82 articles: 13 theoretical studies, 4 literature reviews, and 65 empirical analyses. Of the empirical studies, only 5 use qualitative research methods. In terms of the types of firms represented, 47 studies offer comparative investigations of family and non-family firms' innovative behaviors. The first article was published in 2003, and only in the past 2 years has family firm innovation received truly significant attention, with 22 and 21 articles published in 2015 and 2016, respectively. The review of these articles that follow in this chapter distinguishes between theoretical and empirical articles; it further classifies empirical articles according to whether they address (1) the relationship between family firms and innovation inputs, (2) the relationship between family firms and innovation outputs, or (3) other topics related to family firms' innovative activities.

3.3   Family Firms and Innovation Inputs

Decisions regarding R&D investments have attracted significant research attention across a wide array of disciplines. These typically long-term investments encompass high degrees of uncertainty, and their outcomes depend significantly on market- and technology-based complementary assets (Teece 1986), such that R&D investments often result in protracted periods marked by negative cash flows, which may damage firms' short-term performance.

Among the lines of research that have investigated how R&D investment decisions are managed in the context of family business, one interesting approach leverages an agency lens, with the notion that R&D investments represent managerial decisions that emphasize the differences between shareholders' (principals') and managers' (agents') interests. For example, Munari et al. (2010) suggest that a traditional agency theoretical framework supports predictions of a positive relationship between the share of family ownership and investments in R&D activities, for several reasons. First, family members likely have enduring ties to their companies, reflecting the prospect of generational succession. This condition should drive them to invest "patient financial capital" (e.g., Habbershon and Williams 1999) to support long-term projects, including R&D, that increase the chances of competitive advantages and support the development of a healthy organization. Second, family-owned firms often are characterized by an overlap between owners and managers, so the Type 1 agency costs due to information asymmetry and moral hazard should be low (Fama and Jensen 1983), and beneficial R&D activities likely receive appropriate levels of support and investment.

However, agency perspectives on the role of family ownership also can account for risk-taking behaviors by different ownership entities (Fama 1980; Fama and Jensen 1983) and for the existence of principal–principal conflicts. If the family represents a risk-averse type of owner (Naldi et al. 2007), it could have a negative influence on R&D investments. In particular, when families invest a substantial part of their wealth in the company, they have an incentive to maintain the *status quo* (Morck and Yeung 2003), even if it leads to economically irrational conduct that could be detrimental to non-family shareholders (Type 2 agency costs). Moreover, family firms are deeply concerned about their capital preservation, to the extent that they may seek to maintain a prudential financial policy and avoid the capital market as a source of equity. The resulting capital constraints leave them less able to devote substantial resources to ambitious R&D projects, compared with their non-family counterparts (Thomsen and Pedersen 2000). These arguments, suggesting that family owners favor more conservative investments (Morck and Yeung 2003) rather than risky R&D projects, receive support from an empirical analysis in Western Europe that demonstrates that the presence of family owners reduces the level of investment in R&D (Munari et al. 2010). This negative association between family involvement and R&D intensity receives further support from Muñoz-Bullón and Sanchez-Bueno (2011), who consider other potential factors underlying this relationship, such as non-traditional agency costs stemming from altruism toward specific family members or a lack of technical competence that results from excessive leveraging of non-professional human capital. Such trends might reduce the family's effective monitoring and discipline, to the detriment of their appropriate R&D decision making.

Generally then, empirical support for a lower R&D investment by family-controlled firms is fairly widespread (Chen and Hsu 2009; Chrisman and Patel 2012; Patel and Chrisman 2014; Anderson et al. 2012; Nieto et al. 2015; Classen et al. 2014; Duran et al. 2016;

Broekaert et al. 2016). This evidence also has found ample theoretical support in the behavioral agency model (BAM) and the central role it assigns to the family's socio-emotional wealth (SEW). That is, family firms' aversion to the potential loss of their SEW leads them to avoid R&D investments that could modify their current status, strategic positioning, or assets (Gómez-Mejía et al. 2007; Berrone et al. 2010).

Still, this generalized finding that family firms invest less in R&D than non-family firms has been challenged by several empirical observations (Llach and Nordqvist 2010; Chrisman and Patel 2012; Kotlar et al. 2014; Sciascia et al. 2015). For example, family firms increase their R&D investments when they face threats to their survival (Chrisman and Patel 2012) or profitability (Kotlar et al. 2014; Gomez-Mejia et al. 2014). In such cases, economic rationality and noneconomic goals converge in driving firm decisions; in fact, family firms actively seek to overcome performance hazards and avoid firm failure, which would cause the irreversible loss of SEW derived from firm control (Chrisman and Patel 2012). Similarly, family firms exhibit higher variability in their R&D investment decisions compared to non-family firms. Their desire to preserve their current SEW conflicts with their pursuit of long-term objectives, such as transgenerational control. Thus, the relative importance different family firms may ascribe to short- vs. long-term goals is the source of significant heterogeneity in their R&D conduct (Chrisman and Patel 2012).

A family firm's economic performance might influence not only the level but also the nature of its R&D activities. When performance drops below the family's expected targets, R&D investments get allocated to explorative projects, which offer more substantial opportunities to correct current performance trends (Patel and Chrisman 2014). These findings, which have been confirmed also in the context of small- and medium-sized enterprises (SMEs) (Sciascia et al. 2015), emphasize the diverse goals that animate families and confirm that family firms' actions to preserve their SEW do not necessarily conflict with the canonical economic goals of growth and survival.

Agency-based studies also have attempted to account for the growing evidence that suggests family firms are highly heterogeneous. For example, the R&D behavior of firms that are merely owned by a family and those that are also managed by family members could differ, so Block (2012) distinguishes the ownership and management dimensions. The resulting empirical analysis suggests that though family management does not appear to exert any significant influence on R&D investments and productivity, ownership critically matters. This author in turn separates family from lone founder firms (i.e., companies in which one of the founders serves as an executive/major shareholder, but no relatives are involved in firm ownership or management; Miller et al. 2007). Lone founders exhibit strong entrepreneurial characteristics and are likely to be highly cognizant of their company's business, due to their early involvement in its development. These conditions tend to reduce R&D-related agency costs, because they ensure more appropriate monitoring of R&D activities and the incentives they generate. Accordingly, in lone founder firms, both R&D spending and its productivity tend to be higher than in family firms (Block 2012). Miller et al. (2011) confirm both the conservative approach preferred by family firms, which results in lower R&D intensity, and the positive effect of lone founders on R&D investments, though their theoretical framework mainly emphasizes the social context of ownership; for lone founders, the social context mainly consists of venture capitalists, partners, investors, and other stakeholders, not family members. Thus, lone founders are more emotionally detached and financially driven, because they operate independently from a "familial" approach to business.

Although Schmid et al. (2014) confirm a negative relationship between family control and R&D investments, they also argue and show empirically that in firms with active family management, R&D investments tend to be significantly higher, but this positive effect can be ascribed mainly to the founders, not subsequent generations. Family managers are less sensitive to employment risk and incentives for maximizing their external reputation (Narayanan 1985; Berk et al. 2010); particularly, founders often possess superior technological skills that can help them make appropriate decisions. These conditions may explain the positive effect of a founder's active involvement in managerial decisions. According with the authors (Schmid et al. 2014), the results regarding the significant effect of family management, which contrast with previous literature (Block 2012), might be understood in light of the potentially greater opacity of family firms. Data on R&D personnel ratios, which are used in this study instead of traditional accounting measures of R&D expenditures, offer one means to control for family firms' general tendency to underreport their expenditures (Schmid et al. 2014).

To extend the investigation of the multidimensional effects of the family on R&D investments, Matzler et al. (2015) distinguish among family ownership, management, and governance. According to their empirical analysis, the share of equity owned by the family does not influence R&D intensity, but both family management and governance reduce the level of R&D activities performed within the firm. On the one hand, family managers likely make decisions that are not totally economically rational, to ensure that the family maintains control over the firm, even when such decisions threaten potentially productive R&D projects. On the other hand, in firms whose governance entails the presence of a family-dependent board of directors, familial interests will be better protected, to the detriment of R&D investments that could disrupt the firm's cash flow. Finally, Sirmon et al. (2008) consider the role of family influence, which they define as a combination of family ownership and family managerial presence that ensures the family can offer significant guidance regarding the firm's strategic actions, without exercising unilateral jurisdiction. According to these authors, this desirable condition coalesces some benefits of family involvement (e.g., family-specific resources such as patient capital) while simultaneously preventing its disadvantages (e.g., family entrenchment, altruism, nepotism, myopic traditions). Therefore, they find that family-influenced firms can deal with threats of imitation more effectively than other firms, such as by ensuring appropriate levels of R&D investments to safeguard and expand the firm's competitive advantages.

Beyond agency theory, scholars have worked to unravel the relationship between the family and the firm's R&D investment using stewardship theory (Davis et al. 1997), particularly in analyses of idiosyncratic, emerging market contexts. According to a stewardship view, the wealth, career opportunities, and reputation of family members are inherently connected to the company's performance. Thus, the family has strong incentives to support the long-term interests of the company and its shareholders (Le Breton-Miller and Miller 2009). Inspired by this theoretical approach, Ashwin et al. (2015) study the Indian pharmaceutical industry and find that family ownership, management, and CEO duality all increase the firm's R&D investment. Family owners' focus on the firm's longevity and continuity drives them to prefer investments in innovation efforts, rather than short-term projects. Similarly, family managers, as stewards of the firm, are less prone to engage in opportunistic behaviors, so they seek to mitigate the risk associated with R&D investments by promoting a trust-based, transparent social environment. Finally, overlap between a CEO role and the chairperson position should reinforce the former's positive influence on R&D investments, in light of the increased decision-making effectiveness produced by this special family management scenario. With similar reasoning, Yoo and Sung (2015) investigate the de facto control exerted by a family over the firm's decisions, pointing to situations in which the family is the single largest shareholder and simultaneously holds the most seats on the board. They document a positive effect on R&D investments, particularly when the firm's growth opportunities are limited. During the Asian financial crisis in the late 1990s for example, the social responsibility experienced by Korean families spurred them to invest significantly in value-creating activities such as R&D. Similarly, Singh and Gaur (2013) argue that in emerging countries, family owners act as managers and take care of the firm's strategic activities, which minimizes traditional agency costs, such that family ownership is positively associated with R&D investments.

3.4 Family Firms and Innovation Outputs

The idiosyncratic configuration of family resources, incentives, and goals in turn influences innovation outputs. Although investing in innovation emerges from a strategic choice, the outcomes of the innovative activities also depend on the firm's knowledge, structure, processes, and capabilities (Matzler et al. 2015). These factors significantly influence the way the innovation inputs, which result from an explicit managerial choice, get transformed into actual innovative products or processes. The understanding of the actual relation between the role of the family and the firm's innovation outcomes thus constitutes a complex challenge.

Many studies of innovation outputs in family firms rely on the same arguments that prevail in the literature on innovation inputs. They reflect the characteristics traditionally ascribed to family owners, including their risk aversion, the tight control they maintain over equity, their limited ability to fund innovation projects, and their readiness to preserve the status quo. This strand of literature emphasizes the role of human capital, suggesting that family firms are unwilling to recruit external employees, which limits the pool of human resources. This approach to the composition of the workforce likely reduces the firm's access to new ideas and skills, and the resulting closed innovation environment likely has a negative influence on the firm's innovation performance. Scholars acknowledge the negative influence of paradoxical thinking in family firm innovation too, suggesting that the well-known tensions that characterize family firms (e.g., tradition versus change, control versus autonomy, liquidity versus growth) may generate organizational pressures and hinder decision making, thus obstructing innovation (Ingram et al. 2016). Not surprisingly then, several studies document lower innovation performance in family, compared with non-family, firms (Chin et al. 2009; Anderson et al. 2012).

The negative effect of family ownership on innovativeness also has been confirmed in firm succession contexts. Despite some predictions that later family generations can serve as conduits for an innovation-oriented organizational culture and greater professionalism (McConaughy and Philips 1999), research shows that later generations actually have a negative impact on a family firm's innovation (Beck et al. 2011; Memili et al. 2015). For example, Grundström et al. (2012) find that within-family successions are accompanied by low innovation intensity and a focus on incremental advancements; family-specific values seemingly get retained in the succession of a family-owned firm to family members, which facilitates the preservation of the existing frames, business relationships, and approaches to managing innovative activities within the firm. Kammerlander et al. (2015) also suggest that a powerful influence on the relationship between within-family successions and family firms' innovation is the set of stories that get transmitted across generations. Family firms that share their family stories across generations are more likely to introduce innovations than family firms in which the shared stories mainly focus on the founder. In a related finding, Letonja and Duh (2016) suggest that for within-family successions, next-generation innovativeness occurs only when the tacit knowledge transferred by the founder is combined with a substantial infusion of external knowledge, gained independently of the family.

With a slightly different focus, Kraiczy et al. (2015) demonstrate that the positive effect of a CEO's risk-taking propensity on the innovativeness of a firm's new product portfolio is weaker in family firms at later generational stages. They suggest that the founding generation often encourages innovative initiatives to ensure value creation, which leads to a healthy, viable business to transfer to future generations. But as the generations in charge grow farther distant from the founder's generation, they likely exhibit less propensity to engage in risky entrepreneurial projects, perhaps due to their weaker identification with and attachment to the firm and less intense focus on the preservation of SEW.

De Massis et al. (2015a) show that family firms' innovation endeavors tend to be devoted to minor improvements of existing products, rather than path-breaking new product development plans, likely due to their desire to retain firm control, in combination with their limited resource availability that prevents participation in costly experimentation processes. Leveraging an SEW perspective, Block et al. (2013) separate the effects of family ownership versus management on the technological value of a firm's innovation. Beyond their tendency toward low R&D investments, family firms also appear to direct their investments toward projects that are relatively less novel and ambitious. This behavior is consistent with the goal of preserving family control, because breakthrough innovation entails more significant risks and requires substantial financial resources and highly skilled human capital. An empirical analysis supports these theoretical expectations that the negative effect of family management on the technological impact of innovation is stronger than the effect of family owners, because these managers directly influence the firm's agenda. In a consistent finding, Liang et al. (2013) show that family members' involvement in management teams negatively affects the relationship between R&D investments and innovation performance, whereas family involvement in the board has the opposite effect. Along similar lines, Cucculelli et al.

(2016) demonstrate that family-managed firms are less likely to introduce products that deviate from the firm's existing portfolio in terms of industry and technological characteristics. In turn, these firms resist the renewal of their technological competences, unless they are threatened by a financial crisis; in this condition though, they perceive the introduction of new, riskier products as desirable, because it may enable the firm's survival, in accordance with recent extensions of the BAM (Chrisman and Patel 2012).

In the presence of a higher proportion of family versus non-family members in the top management team, the positive relationship between the management's innovation orientation, defined as the support and commitment senior managers exhibit to innovation, and new product portfolio's performance suffers (Kraiczy et al. 2014). Moreover, when family members involved in the top management team also possess significant ownership levels, the positive effect of a CEO's risk-taking propensity on new product portfolio innovativeness grows weaker, because the family's substantial power and pronounced focus on its SEW reduce the CEO's managerial discretion to engage in ambitious product development projects (Kraiczy et al. 2015). But when family firms are managed by non-family managers, their innovation strategy is more ambitious and targeted toward technological excellence and breakthrough innovations, because the external executives are not concerned about the preservation of the firm's SEW (Lazzarotti and Pellegrini (2015). Thus, Dieguez-Soto et al. (2016) suggest that family firms that rely on professional, external managers achieve greater innovation in both their product and process technologies.

The negative relationship between family ownership/management and innovation is not unambiguous though. Research that adopts the resource-based view (RBV) or stewardship theory emphasizes the potentially positive effects on innovation performance. For example, family firms can effectively leverage their long-term relationships with key stakeholders, which should increase the circulation of ideas and knowledge (Classen et al. 2014). Shifting the focus of the analysis to a more macro-level, research has demonstrated that regions featuring a higher density of family firms tend to be more successful in terms of patent production (Block and Spieger 2013). Some studies thus show that family-owned firms have an equivalent (Classen et al. 2014) or even higher (Gudmundson et al. 2003) ability to introduce new products and services, while they also are more effective at attaining process innovation (Classen et al. 2014) than non-family firms. The influence of family-specific resources and capabilities on innovation outputs may be stronger when the family is actively involved in the management and governance of the firm (Matzler et al. 2015), rather than mere ownership (Banno 2016). Their empirical test shows that family management has a positive influence on both a firm's patent production and the quality of its innovations, as captured by the citations these patents receive, while family governance only seems to affect patent production (Matzler et al. 2015). Theoretically, these findings reflect the notion that family managers are less exposed to job security pressures and less concerned about short-term performance, but they are more embedded in the firm's business network and more central in its communication flows. These conditions generate a powerful combination of lower agency costs and productive familial resources, which increases family managers' commitment to firm success, well beyond their managerial responsibilities, and facilitates flexible decision making. In turn, these mechanisms promote innovation performance (see also Lopez-Fernandez et al. 2016a). In terms of governance, because family involvement in the board is mainly meant to advise managers and ensure the perpetuation of SEW (Corbetta and Salvato 2004), it may be associated with wise innovation decisions (Matzler et al. 2015). However, this finding is controversial, in that some research identifies an opposite effect (Banno 2016).

The innovation performance of family firms also might depend on the context, such as developed versus emerging countries. Focusing on India, Lodh et al. (2014) recommend combining agency theory with institutional perspectives to understand why, when judicial, regulatory, and governance institutions and systems are weak, firms with very concentrated family ownership may benefit from their substantial control over managerial decisions. In such contexts, family firms also may enjoy rich social and political capital and significant access to financial and raw materials, which enable them to focus on performance goals rather than on expropriating minority shareholders. Family ownership in India thus tends to have a positive influence on innovation productivity; in a similar vein, Hsieh et al. (2010) find that in Taiwan, family ties positively affect firms' patent production.

In a recent study, Duran et al. (2016) propose an interesting view on innovation that attempts to reconcile evidence that shows that family firms tend to underinvest in R&D with evidence that family firms are among the most productive innovators worldwide. Leveraging arguments pertaining to control, wealth concentration, and non-financial goals in family firms, these authors suggest that family firms may have limited financial resources and a marked preference for low-uncertainty investments, such that they allocate fewer resources to R&D activities than do non-family firms, but they simultaneously engage in idiosyncratic conversion processes to deal with these innovation inputs. In particular, family firms appear more efficient than non-family firms in exploiting their own limited R&D investments to generate innovation outputs. This capability—making the best of what they have—may arise from family firms' resource orchestration capability and ability to deploy a combination of firm-specific tacit knowledge and valuable information obtained through privileged access to relationships with reliable external partners. However, the prediction regarding the greater efficiency of family firms in converting innovation inputs into outputs has been both validated (Matzler et al. 2015; Broekaert et al. 2016) and refuted (Dieguez-Soto et al. 2016) in empirical literature, so it requires some caution.

3.5 Theories of Innovation in Family Firms

Theoretical contributions to this topic have explored a wide range of research questions, producing useful insights in the form of (1) new theoretical frameworks for understanding the topic, (2) more in-depth explorations of specific topics, or (3) new classifications. For example, Carnes and Ireland (2013) take up the challenge of identifying the mechanisms through which the RBV can help explain the influence of family involvement in different aspects of a firm's innovation. Starting with the well-supported idea that a firm's endowment with valuable and unique resources and capabilities must be supplemented by appropriate resource management processes to generate competitive advantages, these authors investigate the processes by which critical family firm resources—namely their familiness—influence their innovation outcomes. They focus on resource bundling or the process through which managers recombine and transform firm resources into value-creating capabilities by stabilizing, enriching, and pioneering mechanisms. The subprocesses involved in enriching and pioneering resources both relate positively to innovation, because they support the expansion and renewal, respectively, of the firm's existing capabilities to produce incremental and radical advancements. However, the stabilizing process has a negative impact on innovation, because it is oriented toward the preservation of the firm's current capabilities. Families enjoy trusted social relationships and possess extensive tacit knowledge about key assets, which may facilitate their enriching processes. They also are strongly committed to the perpetuation of their current strategies, particularly if they have proven successful in the past. Thus, familiness likely supports stabilizing processes, rather than facilitating pioneering. In summary, familiness, which is the most distinctive and casually ambiguous family firm resource, spawns heterogeneous (positive and negative) influences on innovation, through the specific resource bundling mechanisms it activates. These findings help explain some of the mixed results in empirical literature on innovation in family firms.

Other scholars theorize about the role of additional familial resources. Penney and Combs (2013) extend the theoretical model by Carnes and Ireland (2013) by adding dimensions from the so-called circumplex model of family systems, according to which families represent problem-solving units (Olson et al. 1989). Focusing on cohesion, or the degree of emotional attachment linking family members, and flexibility, which represents variability in family members' roles and leadership degrees, these authors argue that different combinations define the family firm's bundling processes and thus their innovative outcomes, such that they can explain why different families achieve varying levels of innovation performance. Extending this line of research, De Clercq and Belausteguigoitia (2015) argue that two aspects of familiness—the approach to conflict management and social capital—largely determine whether the involvement of different family generations in strategic decision making favors innovation. For Bennedsen and Foss (2015), several family-specific assets are critical to innovation, including the family firm's name, values, reputation, and embeddedness in relevant social networks. The integration of these resources potentially facilitates engagement in productive innovation processes, though over time, these assets also might become rigidities that constrain, instead of stimulating, family firms' innovation. To counter this risk, innovation may need to be institutionalized in the family firm, using mechanisms such as corporate governance or individual incentives.

Miller et al. (2015) agree that family firms' resources can be double-edged swords. Some family-specific assets and objectives encourage innovation; others hinder firm renewal. They classify socio-emotional goals using two categories: "family-centric" and focused on the private interests of the family and its members, which often entails nepotism, parental altruism, and managerial entrenchment, or "firm-centric" and oriented toward the long-term survival of a healthy and successful organization, by means of professionalism, openness to non-family members, and appropriate corporate governance mechanisms. By matching these two categories to different environmental conditions, these authors develop a typology of approaches to innovation. For their classification of family firms' innovation strategies, Li and Daspit (2016) leverage literature on family governance and SEW. In addition to descriptive power, these studies contribute to the field by demonstrating the vast heterogeneity and complexity of family firms, which cannot be addressed with rigid schemas that tend to be too simplistic to capture the multifaceted phenomena.

König et al. (2013) propose that family firms differ from non-family firms when it comes to the adoption of discontinuous technologies. In their pursuit of continuity, command, community, and connections (Miller and Le Breton-Miller 2005), family firms operate under unique constraints that may mitigate some inertial pressures while simultaneously accentuating other forms of resistance. Consider, for example, that family firms may have less formalized organizational structures than non-family firms, but they are also more bonded to noneconomic values. These differences engender an idiosyncratic dilemma and thus an entirely distinct discontinuity adoption process, in both temporal and modality aspects. In this respect, the model the authors propose suggests that family firms are slow to recognize the existence and potential of discontinuous technologies but then become faster and more vigorous in the implementation stage than non-family firms (König et al. 2013). This model has been criticized by Chrisman et al. (2015b) though, for its supposedly poor ability to account for the sources of family firms' heterogeneity, such as business and non-business goals, governance modes, and resources. Building on studies inspired by advances in the BAM (Chrisman and Patel 2012) and in contrast with König et al.'s (2013) predictions, they therefore suggest that family firms facing very serious threats to their survival will rapidly recognize opportunities embedded in a discontinuous technology.

Extant frameworks also integrate theories that have not been applied to studies of innovation inputs or outputs. For example, Chrisman et al. (2015a) consider an ability and willingness framework, in which *ability* refers to the family's discretion to act, due to its dominant position within the firm, and *willingness* indicates the family's disposition to participate in distinctive behaviors inspired by its goals and incentives. In an innovation setting, these family-specific features generate a paradox: The family has sufficient legitimacy and power to promote innovation, but due to its noneconomic goals, it is reluctant to alter the status quo through innovative activities. Recent empirical research (Steeger and Hoffmann 2016) identifies a consequence of this paradox: Family firms innovate less than non-family firms, despite their potentially greater ability to innovate. The outcome of this paradox in turn varies with the degree and sources of ability and willingness.

Other studies that adopt an ability–willingness perspective address more specific innovation-related issues. For example, Veider and Matzler (2016) apply the framework to organizational ambidexterity, arguing that family firms' ability to achieve ambidexterity is subject to their willingness to cope with the challenges arising from the idiosyncratic features of their governance systems, resources, and intentions. Focusing on technology adoption, Holt and Daspit (2015) leverage the ability–willingness dichotomy to develop a *readiness for innovation* construct. For family firms to be ready to adopt a technological innovation, both family and non-family members must be both able and willing to internalize that innovation. These conditions in turn depend on various structural and psychological factors, at both individual and collective levels. In a refinement of the ability–willingness model, De Massis et al. (2015b) distinguish between willingness as a disposition, ability as a discretion, and ability as resources. Then, these three sources of heterogeneity should determine the sources of heterogeneity in innovation decisions, such as the direction of the innovation search, the strategic approaches underlying management of the innovation process, and the types of innovation pursued. According to these authors, fit between these two sets of elements would help family firms overcome typical innovation paradoxes.

Finally, De Massis et al. (2016a) combine innovation, dynamic capabilities, and family business literature to propose *innovation through tradition*, that is, a product innovation strategy that relies on temporal search processes and seeks to redeploy extant knowledge to create new value. This strategy clearly matches the historical and organizational context of family firms; it highlights the importance of reinterpreting past knowledge to avoid mistakes and to improve the reliability and acceptance of innovations, particularly in industries in which experience is a useful reference for customers and tradition activates positive emotions. Families enjoy preferential access to past knowledge, through their transgenerational succession, shared history, and values, so their innovation processes may build successfully on tradition to develop new product concepts.

3.6 Other Areas of Investigation into Family Firms' Innovative Activities

Most empirical research admittedly has focused on innovation inputs and outputs, but several scholars consider the broader topic of innovation in family firms. The resulting articles constitute two categories: studies focused on the moderating and mediating effects that influence the relationships between mainly (though not exclusively) family-related variables and innovation-related variables, and studies that address specific areas related to the management of innovation in family firms.

In the first category, a subset of research investigates variables that might alter the relationship between family involvement and innovation. Craig and Moores (2006) note the conditions that foster innovation in family firms and find that the degree of techno-economic uncertainty positively correlates with innovation; family firms seemingly react to risky environments by innovating more. Hsu and Chang (2011) investigate how family-owned firms use behavioral strategic control, a tool that is based on personal and face-to-face contacts. Their results suggest that family ownership is associated with the greater use of behavioral strategic control, which then has positive effects on innovation. Similarly, scholars identify some organizational design decisions (De Massis et al. 2016b) and behavioral approaches (Covin et al. 2016) that support innovation in family firms.

Family involvement variables also might function as potential moderators of the relationship between different aspects of a firm's strategic approach and its innovation performance. For Tsao and Lien (2013), a key point is how family management interacts with internationalization to explain innovation. Craig et al. (2014) consider the effect on innovation of risk-taking and proactivity, across family versus non-family firms. Sanchez-Famoso et al. (2015) also investigate whether the relationship between a family firm's social capital and its innovation is moderated by the level of family ownership; family ownership has been investigated as a moderator of the relationship between market orientation and innovation too (Newman et al. 2016).

Finally, some studies in this first category address the ways that different innovation-related variables influence the relationship between family involvement and firm performance (Kellermanns et al. 2012). They also investigate how different aspects pertaining to the family might interact with the firm's innovation to explain its performance (Spriggs et al. 2013; Hatak et al. 2016).

In the second category of studies, most contributions focus on specific innovation processes and activities, covering topics that range from family firms' reactions to discontinuous technological change (Kammerlander and Ganter 2015), to succession management for innovation purposes (Hauck and Prugl 2015), to involvement in spontaneous innovation activities by employees (Bammens et al. 2015), and to the most critical obstacles that family firms face in their innovation process (Lopez-Fernandez et al. 2016b). Particular emphasis has been devoted to the open innovation strategies of family firms; it appears that family firms are less likely to acquire external knowledge (Kotlar et al. 2013), and when they do, they limit their search to a relatively narrow set of potential partners (Basco and Calabro 2016).

The analysis of this set of articles thus reveals a significant variety of subtopics, theoretical angles, and empirical approaches. Although they advance knowledge of innovation in family firms, these studies also confirm the vast amount of work that remains to be done to gain a clear understanding of this broad, deeply relevant subject. Scholars willing to pursue this objective should realize that the path forward is not without hurdles, as the concluding section outlines.

3.7   Concluding Remarks

This literature review and survey produces several key conclusions that should inform further research. They reflect some of the most important challenges that remain, according to this critical analysis of the existing body of literature.

1. *Ambiguity in empirical literature.* Empirical literature on innovation in family firms is relatively young, but even still, a significant degree of ambiguity has arisen regarding the central relationships investigated, such as those between the different dimensions of family influence (ownership, management, governance) and the critical innovation-related variables of inputs and outputs. This claim requires further specification though. Research into innovation inputs tends to be more unanimous in citing a negative influence of the family firm on the level of R&D investments, compared with the stream on innovation outputs. Yet even in this research subfield, significant disagreement persists regarding the distinct roles of the various dimensions of family influence.

2. *Heterogeneity of family firms.* The heterogeneity of family firms clearly drives the varied results available in empirical literature. The incredible number of governance-, incentive-, and resource-related variables,

especially in combination, gives rise to innumerable, idiosyncratic situations that may heterogeneously affect the firm's innovation behavior.

3. *Operationalization of variables of interest.* The complexity associated with accounting for the wide set of family dimensions that likely influence family firm innovation is accentuated by the empirical challenge of identifying appropriate, homogeneous ways to operationalize these theoretical constructs. This challenge also involves the innovation side of the relationship, making the objective of isolating a set of uncontroversial findings about family firms' innovation even more difficult, because so many indicators have been used in prior literature to capture both family dimensions of ownership, management, and governance and the inputs/outputs of firms' innovative activities.

4. *Context specificity of the results.* Literature on both innovation inputs and outputs in family firms clearly reveals that the relationships depend strongly on the institutional context of the observed firms. This is not surprising; country-specific regulations, governance factors, and the varied set of social, political, and cultural issues have substantial influence over family firms' innovative behavior. Scholars thus should devote sufficient attention to defining the boundary conditions in which the predictions they derive from their own empirical analyses are most likely to hold, or not.

References

Anderson, R. C., Duru, A., & Reeb, D. M. (2012). Investment policy in family controlled firms. *Journal of Banking & Finance, 36*(6), 1744–1758.

Ashwin, A. S., Krishnan, R. T., & George, R. (2015). Family firms in India: Family involvement, innovation and agency and stewardship behaviors. *Asia Pacific Journal of Management, 32*(4), 869–900.

Bammens, Y., Notelaers, G., & Van Gils, A. (2015). Implications of family business employment for employees' innovative work involvement. *Family Business Review, 28*(2), 123–144.

Banno, M. (2016). Propensity to patent by family firms. *Journal of Family Business Strategy, 7*(4), 238–248.

Basco, R., & Calabro, A. (2016). Open innovation search strategies in family and non-family SMEs evidence from a natural resource-based cluster in Chile. *Academia-Revista Latinoamericana de Administracion, 29*(3), 279–302.

Beck, L., Janssens, W., Debruyne, M., & Lommelen, T. (2011). A study of the relationships between generation, market orientation, and innovation in family firms. *Family Business Review, 24*(3), 252–272.

Berk, J. B., Stanton, R., & Zechner, J. (2010). Human capital, bankruptcy, and capital structure. *The Journal of Finance, 65*(3), 891–926.

Bennedsen, M., & Foss, N. (2015). Family assets and liabilities in the innovation process. *California Management Review, 58*(1), 65–81.

Berrone, P., Cruz, C., Gomez-Mejia, L. R., & Larraza-kintana, M. (2010). Socioemotional wealth and corporate responses to institutional pressures: Do family-controlled firms pollute less? *Administrative Science Quarterly, 55*(1), 82–113.

Block, J. H. (2012). R&D investments in family and founder firms: An agency perspective. *Journal of Business Venturing, 27*(2), 248–265.

Block, J., Miller, D., Jaskiewicz, P., & Spiegel, F. (2013). Economic and technological importance of innovations in large family and founder firms: An analysis of patent data. *Family Business Review, 26*(2), 180–199.

Block, J. H., & Spiegel, F. (2013). Family firm density and regional innovation output: An exploratory analysis. *Journal of Family Business Strategy, 4*(4), 270–280.

Broekaert, W., Andries, P., & Debackere, K. (2016). Innovation processes in family firms: The relevance of organizational flexibility. *Small Business Economics, 47*(3), 771–785.

Bushee, B. J. (1998). The influence of institutional investors on myopic R&D investment behavior. *Accounting Review, 73*(3), 305–333.

Carnes, C. M., & Ireland, R. D. (2013). Familiness and innovation: Resource bundling as the missing link. *Entrepreneurship Theory and Practice, 37*(6), 1399–1419.

Chen, H. L., & Hsu, W. T. (2009). Family ownership, board independence, and R&D investment. *Family Business Review, 22*(4), 347–362.

Chin, C. L., Chen, Y. J., Kleinman, G., & Lee, P. (2009). Corporate ownership structure and innovation: Evidence from Taiwan's electronics industry. *Journal of Accounting, Auditing & Finance, 24*(1), 145–175.

Choi, S. B., Park, B. I., & Hong, P. (2012). Does ownership structure matter for firm technological innovation performance? The case of Korean firms. *Corporate Governance: An International Review, 20*(3), 267–288.

Chrisman, J. J., & Patel, P. C. (2012). Variations in R&D investments of family and nonfamily firms: Behavioral agency and myopic loss aversion perspectives. *Academy of Management Journal, 55*(4), 976–997.

Chrisman, J. J., Chua, J. H., De Massis, A., Frattini, F., & Wright, M. (2015a). The ability and willingness paradox in family firm innovation. *Journal of Product Innovation Management, 32*(3), 310–318.

Chrisman, J. J., Fang, H. Q., Kotlar, J., & De Massis, A. (2015b). A note on family influence and the adoption of discontinuous technologies in family firms. *Journal of Product Innovation Management, 32*(3), 384–388.

Christensen, C. (1997). *The innovator's dilemma: When new technologies cause great firms to fall*. Boston, MA: Harvard Business School Press.

Classen, N., Carree, M., Van Gils, A., & Peters, B. (2014). Innovation in family and non-family SMEs: An exploratory analysis. *Small Business Economics, 42*(3), 595–609.

Corbetta, G., & Salvato, C. A. (2004). The board of directors in family firms: one size fits all? *Family Business Review, 17*(2), 119–134.

Coriat, B., & Weinstein, O. (2002). Organizations, firms and institutions in the generation of innovation. *Research Policy, 31*(2), 273–290.

Covin, J. G., Eggers, F., Kraus, S., Cheng, C. F., & Chang, M. L. (2016). Marketing-related resources and radical innovativeness in family and non-family firms: A configurational approach. *Journal of Business Research, 69*(12), 5620–5627.

Craig, J. B., Pohjola, M., Kraus, S., & Jensen, S. H. (2014). Exploring relationships among proactiveness, risk-taking and innovation output in family and non-family firms. *Creativity and Innovation Management, 23*(2), 199–210.

Craig, J. B. L., & Moores, K. (2006). A 10-year longitudinal investigation of strategy, systems, and environment on innovation in family firms. *Family Business Review, 19*(1), 1–10.

Cucculelli, M., Le Breton-Miller, I., & Miller, D. (2016). Product innovation, firm renewal and family governance. *Journal of Family Business Strategy, 7*(2), 90–104.

Davis, J. H., Schoorman, F. D., & Donaldson, L. (1997). Toward a stewardship theory of management. *Academy of Management Review, 22*(1), 20–47.

De Clercq, D., & Belausteguigoitia, I. (2015). Intergenerational strategy involvement and family firms' innovation pursuits: The critical roles of conflict management and social capital. *Journal of Family Business Strategy, 6*(3), 178–189.

De Massis, A., Di Minin, A., & Frattini, F. (2015a). Family-driven innovation: Resolving the paradox in family firms. *California Management Review, 58*(1), 5–19.

De Massis, A., Frattini, F., Pizzurno, E., & Cassia, L. (2015b). Product innovation in family versus nonfamily firms: An exploratory analysis. *Journal of Small Business Management, 53*(1), 1–36.

De Massis, A., Frattini, F., Kotlar, J., Petruzzelli, A. M., & Wright, M. (2016a). Innovation through tradition: Lessons from innovative family businesses and directions for future research. *Academy of Management Perspectives, 30*(1), 93–116.

De Massis, A., Kotlar, J., Frattini, F., Chrisman, J. J., & Nordqvist, M. (2016b). Family governance at work: Organizing for new product development in family SMEs. *Family Business Review, 29*(2), 189–213.

Dieguez-Soto, J., Manzaneque, M., & Rojo-Ramirez, A. A. (2016). Technological innovation inputs, outputs, and performance: The moderating role of family involvement in management. *Family Business Review, 29*(3), 327–346.

Drucker, P. F. (1985). The discipline of innovation. *Harvard Business Review, 63*(3), 67–72.

Duran, P., Kammerlander, N., Van Essen, M., & Zellweger, T. (2016). Doing more with less: Innovation input and output in family firms. *Academy of Management Journal, 59*(4), 1224–1264.

Fama, E. F. (1980). Agency problems and the theory of the firm. *Journal of Political Economy, 88*(2), 288–307.

Fama, E. F., & Jensen, M. C. (1983). Separation of ownership and control. *The Journal of Law and Economics, 26*(2), 301–325.

Gomez-Mejia, L. R., Campbell, J. T., Martin, G., Hoskisson, R. E., Makri, M., & Sirmon, D. G. (2014). Socioemotional wealth as a mixed gamble: revisiting family firm R&D investments with the behavioral agency model. *Entrepreneurship Theory and Practice, 38*(6), 1351–1374.

Gómez-Mejía, L. R., Haynes, K. T., Núñez-Nickel, M., & Monyano-Fuentes, H. (2007). Socioemotional wealth and business risk in family-controlled firms: Evidence from Spanish olive oil mills. *Administrative Science Quarterly, 52*, 106–137.

Grundstrom, C., Oberg, C., & Ronnback, A. O. (2012). Family-owned manufacturing SMEs and innovativeness: A comparison between within-family successions and external takeovers. *Journal of Family Business Strategy, 3*(3), 162–173.

Gudmundson, D., Tower, C. B., & Hartman, E. A. (2003). Innovation in small businesses: Culture and ownership structure do matter. *Journal of Developmental Entrepreneurship, 8*(1), 1–17.

Habbershon, T. G., & Williams, M. L. (1999). A resource-based framework for assessing the strategic advantages of family firms. *Family Business Review, 12*(1), 1–25.

Hall, P. A., & Soskice, D. (Eds.). (2001). *Varieties of capitalism: The institutional foundations of comparative advantage.* OUP Oxford.

Hatak, I., Kautonen, T., Fink, M., & Kansikas, J. (2016). Innovativeness and family-firm performance: The moderating effect of family commitment. *Technological Forecasting and Social Change, 102,* 120–131.

Hauck, J., & Prugl, R. (2015). Innovation activities during intra-family leadership succession in family firms: An empirical study from a socioemotional wealth perspective. *Journal of Family Business Strategy, 6*(2), 104–118.

Holt, D. T., & Daspit, J. J. (2015). Diagnosing innovation readiness in family firms. *California Management Review, 58*(1), 82–96.

Hsieh, T. J., Yeh, R. S., & Chen, Y. J. (2010). Business group characteristics and affiliated firm innovation: The case of Taiwan. *Industrial Marketing Management, 39*(4), 560–570.

Hsu, L. C., & Chang, H. C. (2011). The role of behavioral strategic controls in family firm innovation. *Industry and Innovation, 18*(7), 709–727.

Ingram, A. E., Lewis, M. W., Barton, S., & Gartner, W. B. (2016). Paradoxes and innovation in family firms: The role of paradoxical thinking. *Entrepreneurship Theory and Practice, 40*(1), 161–176.

Kammerlander, N., Dessi, C., Bird, M., Floris, M., & Murru, A. (2015). The impact of shared stories on family firm innovation: A multicase study. *Family Business Review, 28*(4), 332–354.

Kammerlander, N., & Ganter, M. (2015). An attention-based view of family firm adaptation to discontinuous technological change: Exploring the role of family CEOs' noneconomic goals. *Journal of Product Innovation Management, 32*(3), 361–383.

Kellermanns, F. W., Eddleston, K. A., Sarathy, R., & Murphy, F. (2012). Innovativeness in family firms: A family influence perspective. *Small Business Economics, 38*(1), 85–101.

Klang, D., Wallnöfer, M., & Hacklin, F. (2014). The business model paradox: A systematic review and exploration of antecedents. *International Journal of Management Reviews, 16*(4), 454–478.

Konig, A., Kammerlander, N., & Enders, A. (2013). The family innovator's dilemma: How family influence affects the adoption of discontinuous technologies by incumbent firms. *Academy of Management Review, 38*(3), 418–441.

Kotlar, J., De Massis, A., Frattini, F., Bianchi, M., & Fang, H. Q. (2013). Technology acquisition in family and nonfamily firms: A longitudinal analysis of Spanish manufacturing firms. *Journal of Product Innovation Management, 30*(6), 1073–1088.

Kotlar, J., Fang, H. Q., De Massis, A., & Frattini, F. (2014). Profitability goals, control goals, and the R&D investment decisions of family and nonfamily firms. *Journal of Product Innovation Management, 31*(6), 1128–1145.

Kraiczy, N. D., Hack, A., & Kellermanns, F. W. (2014). New product portfolio Performance in family firms. *Journal of Business Research, 67*(6), 1065–1073.

Kraiczy, N. D., Hack, A., & Kellermanns, F. W. (2015). What makes a family firm innovative? CEO risk-taking propensity and the organizational context of family firms. *Journal of Product Innovation Management, 32*(3), 334–348.

Lawless, M. W., & Anderson, P. C. (1996). Generational technological change: Effects of innovation and local rivalry on performance. *Academy of Management Journal, 39*(5), 1185–1217.

Lazzarotti, V., & Pellegrini, L. (2015). An explorative study on family firms and open innovation breadth: Do non-family managers make the difference? *European Journal of International Management, 9*(2), 179–200.

Le Breton-Miller, I., Miller, D., & Lester, R. H. (2011). Stewardship or agency? A social embeddedness reconciliation of conduct and performance in public family businesses. *Organization Science, 22*(3), 704–721.

Lee, P. M., & O'Neill, H. M. (2003). Ownership structures and R&D investments of US and Japanese firms: Agency and stewardship perspectives. *Academy of Management Journal, 46*(2), 212–225.

Letonja, M., & Duh, M. (2016). Knowledge transfer in family businesses and its effects on the innovativeness of the next family generation. *Knowledge Management Research & Practice, 14*(2), 213–224.

Li, Z. H., & Daspit, J. J. (2016). Understanding family firm innovation heterogeneity: A typology of family governance and socioemotional wealth intentions. *Journal of Family Business Management, 6*(2), 103–121.

Liang, Q., Li, X. C., Yang, X. R., Lin, D. M., & Zheng, D. H. (2013). How does family involvement affect innovation in China? *Asia Pacific Journal of Management, 30*(3), 677–695.

Lieberman, M. B., & Montgomery, D. B. (1988). First-mover advantages. *Strategic Management Journal, 9*(S1), 41–58.

Le Breton-Miller, I., & Miller, D. (2009). Agency vs. Stewardship in Public Family Firms: A Social Embeddedness Reconciliation. *Entrepreneurship Theory and Practice, 33*(6), 1169–1191.

Llach, J., & Nordqvist, M. (2010). Innovation in family and non-family businesses: A resource perspective. *International Journal of Entrepreneurial Venturing, 2*(3–4), 381–399.

Lodh, S., Nandy, M., & Chen, J. (2014). Innovation and family ownership: Empirical evidence from india. *Corporate Governance—An International Review, 22*(1), 4–23.

Lopez-Fernandez, M. C., Serrano-Bedia, A. M., & Gomez-Lopez, R. (2016a). Determinants of innovation decision in small and medium-sized family enterprises. *Journal of Small Business and Enterprise Development, 23*(2), 408–427.

Lopez-Fernandez, M. C., Serrano-Bedia, A. M., & Palma-Ruiz, M. (2016b). What hampers innovation in mexican family firms? *Academia-Revista Latinoamericana de Administracion, 29*(3), 255–278.

Matzler, K., Veider, V., Hautz, J., & Stadler, C. (2015). The impact of family ownership, management, and governance on innovation. *Journal of Product Innovation Management, 32*(3), 319–333.

McConaughy, D. L., & Phillips, G. M. (1999). Founders versus descendants: The profitability, efficiency, growth characteristics and financing in large, public, founding-family-controlled firms. *Family Business Review, 12*(2), 123–131.

Memili, E., Fang, H. C., & Welsh, D. H. B. (2015). Value creation and value appropriation in innovation process in publicly-traded family firms. *Management Decision, 53*(9), 1921–1952.

Miller, D., & Le Breton-Miller, I. (2005). *Managing for the long run: Lessons in competitive advantage from great family businesses*. Boston: Harvard Business Press.

Miller, D., Le Breton-Miller, I., Lester, R. H., & Cannella, A. A. (2007). Are family firms really superior performers? *Journal of Corporate Finance, 13*(5), 829–858.

Miller, D., Wright, M., Le Breton-Miller, I., & Scholes, L. (2015). Resources and innovation in family businesses: The Janus-face of socioemotional preferences. *California Management Review, 58*(1), 20–40.

Morck, R., & Yeung, B. (2003). Agency problems in large family business groups. *Entrepreneurship Theory and Practice, 27*(4), 367–382.

Munari, F., Oriani, R., & Sobrero, M. (2010). The Effects of Owner identity and External Governance Systems on R&D Investments: A Study of Western European Firms. *Research Policy, 39*(8), 1093–1104.

Munoz-Bullon, F., & Sanchez-Bueno, M. J. (2011). The impact of family involvement on the R&D intensity of publicly traded firms. *Family Business Review, 24*(1), 62–70.

Naldi, L., Nordqvist, M., Sjöberg, K., & Wiklund, J. (2007). Entrepreneurial orientation, risk taking, and performance in family firms. *Family Business Review, 20*(1), 33–47.

Narayanan, M. P. (1985). Managerial incentives for short-term results. *The Journal of Finance, 40*(5), 1469–1484.

Newman, A., Prajogo, D., & Atherton, A. (2016). The influence of market orientation on innovation strategies. *Journal of Service Theory and Practice, 26*(1), 72–90.

Nieto, M. J., Santamaria, L., & Fernandez, Z. (2015). Understanding the innovation behavior of family firms. *Journal of Small Business Management, 53*(2), 382–399.

Olson, D. H., McCubbin, H. I., Barnes, H., Larsen, A., Muxen, M., & Wilson, M. (1989). *Families: What makes them work* (2nd ed.). Los Angeles, CA: Sage.

O'Sullivan, M. (2000). The innovative enterprise and corporate governance. *Cambridge Journal of Economics, 24,* 393–416.

Patel, P. C., & Chrisman, J. J. (2014). Risk abatement as a strategy for R&D investments in family firms. *Strategic Management Journal, 35,* 617–627.

Penney, C. R., & Combs, J. G. (2013). Insights from family science: The case of innovation. *Entrepreneurship Theory and Practice, 37*(6), 1421–1427.

Perri, A., & Peruffo, E. (2016). Knowledge spillovers from FDI: A critical review from the international business perspective. *International Journal of Management Reviews, 18*(1), 3–27.

Sanchez-Famoso, V., Iturralde, T., & Maseda, A. (2015). The influence of family and non-family social capital on firm innovation: Exploring the role of family ownership. *European Journal of International Management, 9*(2), 240–262.

Schmid, T., Achleitner, A. K., Ampenberger, M., & Kaserer, C. (2014). Family firms and R&D behavior—New evidence from a large-scale survey. *Research Policy, 43*(1), 233–244.

Schumpeter, J. A. (1934). *The theory of economic development: An inquiry into profits, capital, credit, interest, and the business cycle*. Cambridge, MA: Harvard University Press.

Sciascia, S., Nordqvist, M., Mazzola, P., & Massis, A. (2015). Family ownership and R&D intensity in small- and medium-sized firms. *Journal of Product Innovation Management, 32*(3), 349–360.

Singh, D. A., & Gaur, A. S. (2013). Governance structure, innovation and internationalization: Evidence from India. *Journal of International Management, 19*(3), 300–309.

Sirmon, D. G., Arregle, J. L., Hitt, M. A., & Webb, J. W. (2008). The role of family influence in firms' strategic responses to threat of imitation. *Entrepreneurship Theory and Practice, 32*(6), 979–998.

Spriggs, M., Yu, A., Deeds, D., & Sorenson, R. L. (2013). Too many cooks in the kitchen: Innovative capacity, collaborative network orientation, and performance in small family businesses. *Family Business Review, 26*(1), 32–50.

Steeger, J. H., & Hoffmann, M. (2016). Innovation and family firms: Ability and willingness and German SMEs. *Journal of Family Business Management, 6*(3), 251–269.

Strutzenberger, A., & Ambos, T. C. (2014). Unravelling the subsidiary initiative process: A multilevel approach. *International Journal of Management Reviews, 16*(3), 314–339.

Teece, D. J. (1986). Profiting from technological innovation: Implications for integration, collaboration, licensing and public policy. *Research Policy, 15*(6), 285–305.

Tsao, S. M., & Lien, W. H. (2013). Family management and internationalization: The impact on firm performance and innovation. *Management International Review, 53*(2), 189–213.

Thomsen, S., & Pedersen, T. (2000). Ownership structure and economic performance in the largest European companies. *Strategic Management Journal, 21*(6), 689–705.

Tylecote, A., & Ramirez, P. (2006). Corporate governance and innovation: The UK compared with the US and 'insider' economies. *Research Policy, 35*(1), 160–180.

Veider, V., & Matzler, K. (2016). The ability and willingness of family-controlled firms to arrive at organizational ambidexterity. *Journal of Family Business Strategy, 7*(2), 105–116.

Yoo, T., & Sung, T. (2015). How outside directors facilitate corporate R&D investment? Evidence from large Korean firms. *Journal of Business Research, 68*(6), 1251–1260.

4

Family Firm Innovation in the Global Pharmaceutical Industry

Abstract This chapter describes the global pharmaceutical industry, focusing on its most important characteristics and competitive dynamics. Several features of this industry question the fit of families as an adequate ownership and governance category. This chapter devotes special attention to analyzing the role of innovation, in reference to established literature that has extensively investigated the most important technological trajectories and discontinuities that characterize the industry's evolution, and focuses on the most recent technological dynamics and current organization of innovative activities. Finally, a dedicated focus on the Italian context helps illustrate the country-level environmental conditions that shape the conduct of Italian family firms, as analyzed in Chap. 5.

Keywords Global pharmaceutical industry · Innovation process Italian pharmaceutical industry

© The Author(s) 2017
A. Perri and E. Peruffo, *Family Business and Technological Innovation*,
DOI 10.1007/978-3-319-61596-7_4

4.1   The Global Pharmaceutical Industry: Structure, Competitive Dynamics, and Trends

The pharmaceutical industry traditionally has been regarded as a sector with significant profit potential (Ghemawat 2010). However, its profitability depends on continuous innovation (Roberts 1999) and ongoing R&D spending. Industry growth is driven by the rise and development of novel areas of technological and scientific knowledge (Brusoni et al. 2005). Pharmaceutical firm performance relies on the discovery and commercialization of new chemical entities (De Carolis 2003), and investments in upstream innovative activities are critical for firm survival and growth.

In recent decades, the competitive scenario in which pharmaceutical companies operate has been altered by several changes and challenges (Bruche 2012; Scalera et al. 2015). In the golden age of the industry, big pharmaceutical companies could rely on profits from entirely new chemical entities that became blockbuster drugs (i.e., medicines with sales exceeding 1 billion USD). But finding completely novel, breakthrough molecules has become increasingly challenging (Bruche 2012). Several studies have highlighted a significant decrease in pharmaceutical industry's productivity; the introduction of new and innovative products is becoming both more rare and less economically valuable (Cockburn 2004). Moreover, substantial modifications to the process of drug discovery (i.e., shift from traditional trial-and-error methods to a rational drug design approach) have significantly altered the division of innovative labor in the industry (Gambardella et al. 2000) and made the use of external technology sources increasingly necessary (Cockburn 2004). These changes have prompted pharmaceutical companies to turn to open innovation practices, with the aim of gaining quick access to diverse knowledge inputs.

Moreover, the industry also has experienced the effects of globalization (Gambardella et al. 2000). Established national players have lost significant market share in their domestic markets; at the same time, emerging countries have grown to be not only end markets but also locations for manufacturing and R&D activities (Scalera et al. 2015). Overall, the evolution of the industry is taking place amidst widely varying regulatory regimes that affect the conduct, strategies, and market performance of pharmaceutical companies (Stremersch and Lemmens 2009). One of the most striking examples of regulatory impacts on the competitive dynamics and performance evolution of the industry has been the introduction of the Waxman-Hatch Act in the USA. In 1984, it legalized an abbreviated process for the approval of new drugs that substantially reduced the effort required for generic drug producers to operate in the end market after patent expiration. The new procedures reduced the market entry requirement of generic drugs to a mere demonstration of bioequivalence, thus eliminating previous safety and efficacy testing requirements and exerting salient effects on both the rate of market entry after patent expiration and the nature of competition in the industry (Grabowski et al. 2011; Reiffen and Ward 2005).

More than many other industrial settings, the pharmaceutical industry is characterized by a reliance on science, basic research, and development (Brusoni and Geuna 2003; Brusoni et al. 2005). The discovery of new drugs is significantly inspired by the fundamental knowledge developed in academic organizations and research centers and accessed through scientific publications. Pharmaceutical R&D accounts for about one-fifth of all business R&D spending globally, which, as Petrova (2014) explains, is a remarkable figure for a single industry; thus, the average pharmaceutical company tends to be provided with enough absorptive capacity to deal with even basic scientific advances.

In a context in which technological innovation is key to profit, competitive advantage is strongly influenced by companies' abilities to protect their intellectual property. In the pharmaceutical industry, protection from competitors' attempts to appropriate the outcomes of other firms' innovative efforts is typically granted by filing for patents on new molecules (Gambardella 1995) and obtaining market exclusivity for the resulting drugs (Petrova 2014). Because patent protection is independent from the eligibility of drugs for commercialization, patents are usually sought very early in the drug development process (Gambardella 1995). Normally, market exclusivity shields innovators from imitators' threats of entry into end markets, because it reflects the marketing rights for actual products that are granted to innovating companies upon drug approval (Petrova 2014). Given the benefits that accrue to companies that are able to pioneer specific drugs (Grabowski and Vernon 1987), the speed of the technological innovation process is a key variable (Gambardella 1995). Pioneers not only benefit from temporary monopolies as they dominate the market but also gain significant reputation advantages as they win the innovation race.

Patent protection tends to be stronger in the pharmaceutical industry than in other settings (Levin et al. 1987; Bierly and Chakrabarti 1996), and market exclusivity, in combination with the complementary assets required to operate in the industry (e.g., manufacturing, marketing and distribution capabilities, regulatory relationships), serves as a barrier to entry (Gambardella 1995; Gambardella et al. 2000). However, also in this context, truly innovating companies suffer competitive attacks by imitative firms (Grabowski and Vernon 1987). Several enemies endanger the competitive supremacy of pharmaceutical innovators, particularly in specific national markets and product niches (Gambardella et al. 2000). Following the business model categories proposed by Petrova (2014), it is possible to distinguish between *me-too* drug producers, *follow-on* drug producers, and *generic* drug producers (though me-too and follow-on drugs often are grouped into the same category, namely "imitative drug producers").

Me-too drug producers are not usually deliberate imitators, that is, companies that limit their investments to areas that have already proved successful (DiMasi and Laden 2011); they are more likely to be companies that pursue the development of their own drugs and work on specific therapeutic targets, before the approval of pioneer drugs (DiMasi and Paquette 2004). In other words, they are late entrants that bear the costs of new drug development but do not win the innovation race. Because they need to recoup their R&D investments, they have a significant incentive to enter the market by offering products that are positioned close to pioneer drugs but are formulated differently and perceived differently by end markets. Not surprisingly, for this category of producers, the most critical competitive levers are marketing to physicians and direct-to-consumer advertising (Petrova 2014).

Follow-on drug producers engage in minor refinements of the chemical structure of pioneer medicines but, unlike me-too drug producers, deliberately wait to initiate the process of drug development until after the approval of pioneer drugs. When novel drugs demonstrate clinical weaknesses that are not so severe that they cause the termination of the approval process, these companies seek to exploit opportunities to offer incremental improvements to the original medicines (Petrova 2014). Therefore, follow-on drug producers do not bear the same amount of risk and uncertainty as pioneer innovators and me-too drug producers; they build on therapeutic solutions that have already been tested and proved to be effective. If the market is large and the follow-on drug is sufficiently enhanced, the returns of such a second-mover approach can equal those of a strategy based on the early market entry, at least in the long run (Bottazzi et al. 2001).

Me-too drug producers and follow-on drug producers are both branded competitors of the original innovators of breakthrough medicines. However, the scenario is completely different for generic drug producers. Generic drugs can be marketed at prices that are up to 50% lower than the prices of original branded products (Griliches and Cockburn 1994). They can be introduced after the window of regulatory protection for the original drug terminates, and though they must feature the same active compounds, safety, quality, and dosage structure as the original drug, they do not have to include the same inactive ingredients. The drugs do not have to go through the lengthy approval procedures required for original medicines, but they are expected to demonstrate sufficient degrees of bioequivalence. For obvious reasons, branded manufacturers continue to emphasize attributes of quality and reputation even after their regulatory protection expires to avoid aggressive price-based competition among generics. However, loss of market share is inevitable for branded producers (Grabowski et al. 2011), because the development of the corresponding generic drug market segment is usually very rapid (Bruche 2012).

To summarize, identification of completely new therapies for areas lacking prior treatments has traditionally generated the highest profits, but the refinement of existing medicines is also a significant source of industry prosperity. The dynamics of innovation in the pharmaceutical industry follow either the path of explorative R&D (for the discovery of completely new chemical entities) or exploitative R&D (for the development of imitative products and generics). Yet, as in all businesses, these two approaches are not mutually exclusive; they often coexist within pharmaceutical companies (The Economist 2012).

These characteristics also affect the geography of competition in the global pharmaceutical industry. Traditionally, the global pharmaceutical marketplace has been dominated by North America, Europe, and Japan, which account for more than 80% of global pharmaceutical sales (KPMG 2011). More recently, however, countries such as India, China, and Brazil have emerged not only as buyers but also as producers of drugs. These contexts, once perceived only as end markets in which established pharmaceutical firms could expand their commercialization activities, are attracting both manufacturing and R&D facilities. Under pressure from increasing competition and narrowing profit opportunities, pharmaceutical activities are moving from advanced countries to emerging countries in which end-market demand is poised to exceed traditional market demand in both size and growth rate, and both physical and human resources are available at much lower costs (Scalera et al. 2015).

4.2   Technological Innovation in the Global Pharmaceutical Industry

From a technological standpoint, the global pharmaceutical industry is inherently dynamic (De Carolis 2003). Because industry growth relies strongly on breakthroughs in life sciences and endogenous processes of innovation (Orsenigo 1989; Gambardella 1995), current and future profit opportunities are associated with substantial risks. Although processes have changed over time, drug discovery and development still entails enormous resource investments and lengthy procedures (DiMasi et al. 1991). Moreover, both the architecture of the innovation process and the decision to go to market are highly regulated and monitored (Gambardella 1995).

Unlike many other market-driven sectors, the pharmaceutical industry follows the *technology-push* model, in which value to patients is strongly determined by the rate of progress of scientific advances (Petrova 2014). Historically, two major technological discontinuities ignited the most significant innovations in the industry (Cockburn and Henderson 2000; Petrova 2014). The first occurred in the mid-1970s, when progress in fields such as physiology, enzymology, pharmacology, and molecular biology was boosted by publicly funded research projects. These disciplines fostered greater understanding of the biochemical and molecular mechanisms that underlie many diseases and the causal relationships activated by existing drugs. The second critical technological breakthrough took place between the late 1970s and the early 1980s, when the emergence of biotechnology, and its revolutionary combination with genetic engineering, made more flexible and effective tools and techniques possible (Petrova 2014).

Traditionally, the process of innovation in the pharmaceutical industry has been a sequence of potentially overlapping stages (DiMasi et al. 2003). Proceeding through these stages is not easy; each new compound must be analyzed, and several regulatory requirements must be met. Typically, in the pre-discovery stage, the company isolates a general area of investigation, such as a type of disease (Sharma and Lacey 2004), and accumulates a deep understanding of the area, such as by identifying the causes and potential evolution of the disease. It identifies and tests a biological target (e.g., protein, gene) that could be relevant to the given disease. Subsequently, in the drug discovery stage, it follows a search path for which the expected output is the synthesis of compounds that effectively impact the target and influence the evolution of the disease. In the past, this search stage was realized through a rather inefficient, methodology-based screening of existing molecules. Over time, a more functional methodology, known as *rational drug design,* became widespread. It was a consequence of the molecular biology revolution (Gambardella et al. 2000)—a dramatic procedural change marked by the recognition that new knowledge bases could be pivotal in the process of drug discovery—which led to the emergence of a new technological regime in which the organization of research activities in the industry shifted substantially. The new methodological approach that resulted from this revolution—using advanced analytical, simulation, and predictive methodologies that go beyond traditional, inefficient quasi-random screening—has enabled the design of completely new prototypes of drug molecules.

In the discovery stage, drugs undergo initial testing to optimize their compound structure. In the preclinical R&D stage, two characteristics of molecules are scrutinized: toxicity and potential therapeutic efficacy. This stage usually occurs in the laboratory and may last up to three years. Only a small subset of the initial sample of molecules successfully passes this stage and makes it to clinical trials. Clinical trials test compounds on patients; they usually take six to seven years, though in the past they could last up to ten years (Gambardella 1995). Among other requirements, compounds must prove to offer significant rates of action, acceptable duration of effects, and efficiency in terms of absorption and metabolism. Positive clinical trials do not ensure that drugs are marketable; marketing approval must be granted by the government agency that regulates and controls the pharmaceutical industry (DiMasi et al. 2003), all of which amplifies the uncertainty that characterizes the entire process.

Identifying the optimal breadth of therapeutic research programs to pursue is critical in this industry, because it involves choosing which knowledge bases to develop (Bierly and Chakrabarti 1996). On the one hand, given the complexity of the subject matter, specializing in one therapeutic field allows a company to accumulate enhanced experience and learning; such specialization may significantly boost its effectiveness in that therapeutic area (Chandy et al. 2006) and thereby improve its reputation. On the other hand, such a narrow focus may limit search strategies and generate lock-in mechanisms that impede the development of broader and less risky portfolios of new molecules (De Carolis 2003). Developing more diverse expertise in a few large programs may be particularly beneficial for bigger companies that possess the appropriate resources (financial, organizational, and knowledge-based) to sustain and make the best of technological diversification. Specific knowledge accumulated in one category eventually can be redeployed in other areas, through a process of opportunity recognition and cross-fertilization (Henderson and Cockburn 1996), thus generating scope economies (Petrova 2014).

In terms of knowledge sourcing, inputs from a broad range of domains, such as physiology, molecular biology, biochemistry, analytic and medicinal chemistry, and even more distant disciplines such as information science, have become increasingly relevant to successful innovation in the pharmaceutical industry. Henderson and Cockburn (1994) suggest that the use of diverse disciplines within the pharmaceutical innovation process helps companies enhance their drug research productivity. Similarly, Pisano (1994) confirms the importance of merging different knowledge areas across the phases of a new drug development process. More generally, connection to the scientific network as a source of additional ideas and knowledge inputs is highly beneficial for pharmaceutical companies.

Overall, though the pharmaceutical industry is inherently driven by science and dependent on technology (Petrova 2014), it is important to acknowledge that the ability to exploit returns from innovation successfully requires access to a set of complementary assets, composed of manufacturing capabilities, relationships with physicians, visibility and reputation, and branding capabilities. To be financially prosperous, companies affiliated with the global pharmaceutical industry must be endowed not only with technological skills but also with marketing and regulatory capabilities (Bogner and Thomas 1994; Cool and Schendel 1987; Henderson and Cockburn 1994; Hill and Hansen 1991; Thomas 1990).

4.3 Business Actors in the Global Pharmaceutical Industry: The Role of Family Firms

The foregoing analysis of the salient features of the pharmaceutical industry provides the basis for a better understanding of the business actors that populate its landscape. It is a crucial step in evaluating the fit of family as a possible ownership typology in this industry, which is the key objective of this work.

The discussion suggests that the global pharmaceutical industry traditionally has been an ideal marketplace for large companies. The need for massive investments in R&D (DiMasi et al. 1991), the significance of economies of scale and scope from innovation, especially in drug discovery (Henderson and Cockburn 1996; Chandy et al. 2006), and the relevance of complementary assets and capabilities (Bogner and Thomas 1994; Cool and Schendel 1987; Henderson and Cockburn 1994; Hill and Hansen 1991; Thomas 1990) have driven the growth of pharmaceutical companies. In their quest for blockbuster drugs, so-called Big Pharma firms have evolved into complex organizations, capable of performing all stages of the pharmaceutical innovation process internally, from the creation of basic knowledge, to actual drug marketing and commercialization, to post-market monitoring (Petrova 2014). Their geographical reach has been accordingly high, fueled by waves of mergers, acquisitions, and successive reorganizations. Giant pharmaceutical companies typically are multinationals in search of increasingly wider markets, in which scale and scope economies can be efficiently exploited. At the end of the twentieth century, according to Gambardella et al. (2000), multinational corporations covered about 40–60% of national markets in the advanced world. These firms have adopted a global approach: They retain a significant share of their activities at home, but they also maintain operations across both national and regional borders.

Although large multinational corporations traditionally have dominated the industry, the molecular biology revolution and the resulting rational drug design approach have significantly modified the structure of the sector and the organization of research activities. Starting in the late 1970s, new technological opportunities enabled the entry and diffusion of *biotech* firms (Arora and Gambardella 1990; Powell et al. 1996; Orsenigo et al. 2001), companies that are positioned at the intersection of large, for-profit industrial companies and public research institutions (Cockburn 2006). Biotech firms are small entrepreneurial companies with extraordinary research intensity, particularly in leading-edge technologies; they focus on the discovery and development of biopharmaceuticals (Cockburn 2006). Although they act as propellers of innovation in the modern pharmaceutical industry (Wuyts and Dutta 2008), their limited size, capital, and human resources tend to restrain their abilities to reach the critical mass needed to exploit economies of scale and scope in pharmaceutical innovation and undertake the activities needed to go to market (Petrova 2014). For these reasons, small biotech firms increasingly position themselves as suppliers of specialized procedures and intermediate products for bigger companies (Gambardella et al. 2000). This role enables them to remain focused on the development of forefront technology, while leaving subsequent stages of the pharmaceutical innovation process to large pharmaceutical companies that are better endowed with the economic infrastructure, capital, and market knowledge needed to accomplish actual drug commercialization (Grewal et al. 2008).

This new division of innovative labor in the pharmaceutical industry has not weakened the predominant role of industry incumbents; they continue to play a critical integrative function that connects and combines diverse knowledge bases, and they provide complementary resources and capabilities that other industry organizations, including small biotech firms, do not possess (Gambardella et al. 2000). Established, large pharmaceutical companies have undergone difficult processes of adaptation; they have learned to absorb frontier knowledge from external partners and adopt increasingly networked organizational forms, in which alliances and collaborative agreements with private and public research institutions, universities, and small biotech firms feed their innovation funnels (Arora and Gambardella 1990; Powell et al. 1996; Orsenigo et al. 2001).

In addition to large pharmaceutical companies and small entrepreneurial biotech firms, the pharmaceutical industry is populated by a third category of companies, that is, those that focus on the sale of non-research-intensive medicines. These firms are national companies with markets that tend to be limited to their home countries. They are mostly involved in manufacturing and commercialization activities, and they make limited investments in R&D (Gambardella et al. 2000).

On the basis of this assessment of the pharmaceutical industry and the most important business actors that populate its landscape, a question arises: Is family business a suitable ownership and governance category for this industry? Two conditions seem to make the family–innovation puzzle, discussed in previous chapters, much more complex and empirically relevant. First, the pharmaceutical industry is highly technology intensive: this trait compels organizations that populate this setting to commit massive amounts of resources to innovative processes that are immensely risky and uncertain (Bruche 2012), and thus clashes sharply with family firms' typical conservative approach and risk aversion (Gómez-Mejía et al. 2007). Second, the industry has an inherently global reach: besides colliding with family firms' traditional focus on local markets (Fernández and Nieto 2005), this attribute increases the competitive pressure to which firms are exposed and further emphasizes the importance of pursuing continuous technological upgrading by means of innovative activities. Thus, in such a context, the analysis of the family–innovation relationship is particularly intriguing.

4.4   An Overview of Italian Firms Operating in the Pharmaceutical Industry

Since it began, the Italian pharmaceutical industry's history has been closely linked to individual business initiatives, in which the family dimension of the enterprise has represented the fundamental unit of local economic activities (Frezza 2015). In Italy, the diffusion of large pharmaceutical companies was significantly slowed by the delayed industrialization of several areas of the country, a lower degree of urban development than other countries, and a lack of unified rules and regulations. For these reasons, when the availability of synthetic molecules allowed them to be substituted for handmade materials, Italian firms operating in the industry were not initially able to take advantage of the opportunity.

Taken together, these factors explain the development model of the Italian pharmaceutical industry, which is the result of the evolution of so-called pharmaceutical shops rather than an expansion of the chemical industry. Most domestic pharmaceutical companies originated from the evolution of these shops, which were characterized by secular familiar traditions and focused on local markets. Through the leverage of chemical/pharmaceutical skills, these small companies were able to commercialize various medical products, including pills, ointments, bitters, and elixirs, and gradually grow into bigger, regional companies. Thus, the most successful chemical entities underwent a transition from Galenic pharmaceuticals, which encompassed preparation in the laboratory following either a doctor's prescriptions or the traditional formulas, to pharmaceutical products sold with a recognizable name and brand.

The enhancement of urban and industrial development, coupled with the entrepreneurial spirit of family leaders, contributed to raising awareness of the work-related social sense of the production of medical products. It led the Italian pharmaceutical industry to take its first steps (Frezza 2015). This phenomenon started before Italian unification, mainly in the north of the country. At that time, there was a strong orientation toward trade and industry development; the Napoleonic era and the influence of the French economy stimulated great political and cultural ferment in the north of Italy. In this period, three major Italian pharmaceutical companies emerged from preexisting pharmacies, and all were family firms: Schiapparelli in Turin in 1824; Carlo Erba in Milan in 1837; and Zambeletti in Milan in 1866 (Perugini 2015).

The Italian pharmaceutical industry experienced a flourishing period between the 1800s and the first three decades of the 1900s. In these years, domestic pharmaceutical companies expanded, and pharmaceutical products became increasingly more widespread, especially as tools for public healthcare. The development of the chemical industry gradually brought the country closer to European and international networks in a liberal economy perspective. However, weaknesses of the Italian pharmaceutical industry persisted. Far from following a rational development pattern, either as a by-product of the progress of the domestic chemical industry or as a result of planned strategic investment by the national government, industry growth was propelled mostly by the entrepreneurial spirit of individual industry leaders (Frezza 2015). Domestic companies were numerous but small, scarcely able to compete with large European firms, due to the traditional model that tied pharmaceutical production to the entrepreneurial initiatives of family-based companies. For these firms, it was hard to compete in a sector characterized by the need for major investments in R&D, complementary assets, and economies of scale and scope (Di Masi et al. 1991; Henderson and Cockburn 1996).

The modernization of the Italian industrial system began after World War II; it included the pharmaceutical industry, despite the relevant disparities separating the north from the south of the country, especially in terms of legal frameworks. Protectionism, in both more and less evident ways, helped preserve the national industry. For example, patent protections could not be requested for chemical and pharmaceutical products until 1978; Italian companies therefore tended to replicate foreign-invented products and resist pressure to undertake explorative R&D activities and ambitious technology projects. Unlike companies in other European countries such as France, Italian companies were unable to take the steps toward consolidation and critical mass that might have led to the emergence of larger, global firms that competed with foreign peers.

Taken together, these issues led to the development of an industry characterized by small- and medium-sized enterprises (SMEs). Over time, the limitations of a business sector that was mainly imitative and exploitative, because of its strong ties with family businesses, became clear. That is, it became obvious that the Italian pharmaceutical industry could survive only in a nationally protected context. Not surprisingly, big Italian companies that were willing to enter the international competitive arena started to demand the institution of intellectual property rights to protect their investments in risky research endeavors.

Even after patent protection for pharmaceutical entities was established in 1978, however, R&D investments in the Italian pharmaceutical industry did not experience a clear growth pattern. Reports show a higher propensity of Italian pharmaceutical companies to seek patents in the USA after 1978; however, this pattern seemed to depend more on US patent reforms than on a higher propensity to innovate (Gamba 2016). Challu (1995) reveals how Italian reforms of patent protection caused a decline in the number of new products developed in Italy, and Scherer and Weisburst (1995) identify price controls imposed on the industry as a key reason for this negative effect. Today, the innovation capability of the Italian pharmaceutical industry continues to suffer from limited development of the system for intellectual property protection, at least partially as a result of the relatively low quality of the overall institutional environment at the country level. It has never been able to reach the standards of other countries and still does not recognize the critical importance of intellectual property protection as an incentive for private investment in risky innovation projects.

Although protectionism and the lack of competition have been detrimental, especially given the rise and growth of the Big Pharma model in other countries, several national entrepreneurs have emerged to contribute proactively to the development of the industry. Some Italian companies have even flourished in this period, including Recordati, Angelini, Bertarelli, Golinelli, De Santis, Zambon, Bracco, Dompé, Cavazza, and Della Valle. Their efforts contributed decidedly to the growing international reputation of the Italian pharmaceutical industry (Frezza 2015).

In the 1980s, the Italian pharmaceutical industry began to suffer from the effects of the globalization process. As a result, outsourcing practices were implemented to rationalize costs, which increased the importance of marketing and sales strategies (Gambardella et al. 2000). In previous years, several multinational firms had been attracted to the country, to carry out manufacturing and R&D activities. The Italian government implemented an industrial policy through the state-owned agency, Cassa per il Mezzogiorno (1950–1986), to promote industrial development and attract foreign direct investments (Faini and Schiantarelli 1983). However, in the past 30 years, the failure of these incentives, coupled with the globalization process that made other countries more attractive to outside investors, forced Big Pharma companies to restructure their subsidiaries or even reconsider their presence in Italy.

As in many other European countries, the public sector plays a pivotal role in the Italian pharmaceutical industry. About 75% of total Italian healthcare spending is funded by taxation through Servizio Sanitario Nazionale (SSN), the national health service. However, Italian per capita public expenditures are much lower than in many other European countries, mainly because the prices of pharmaceutical products in Italy tend to be lower (Marketline 2015). Moreover, because the healthcare sector traditionally has been a primary target for cost containment initiatives, the development of spending review policies to reduce public healthcare expenses and the diffusion of generics have strengthened public sector power and reinforced competition among pharmaceutical companies (Farmindustria 2016). A recently implemented public policy limits reimbursement by SSN to the price of generics, to incentivize the prescription of the cheapest drugs; the government has concluded that the social benefits of cheap drugs outweigh the downside of a potential reduction in pharmaceutical firms' ability to invest in R&D.

In addition to controlling pharmaceutical pricing, public authorities determine the approval process of new drugs and the definition of restrictive formularies. The institutional structure of the healthcare service has both national and regional/local levels. Although this system grants greater autonomy in healthcare policy to local bodies, it has resulted in the duplication of bureaucratic restrictions and obligations, with additional regulations often imposed at the regional or local levels. These additional duties and requirements are more numerous than in most other important European markets; they limit access to innovative drugs by requiring the drugs to comply with additional obligations, even after obtaining national formal approval (Farmindustria 2016).

Today, the Italian pharmaceutical industry is characterized by a large number of companies that remain competitive because of highly specialized human resources; they feature high-quality research activities and growing innovation. Italian companies represent 40% of the total industry; 60% are foreign-capital multinational companies with a strong territorial presence in manufacturing activities (Farmaindustria 2016). Although foreign multinationals have restructured their R&D processes around the world to take advantage of open innovation practices, some manufacturing activities are still located in Italy because of human resources specialization, such that the subsidiaries of large multinational corporations still dominate. Of the remaining companies, only a few are controlled by private equity investors; while most are family firms (nearly 40% of the total industry). Among these family firms, there are a number of *national champions*—relatively large

Italian companies with a long tradition in the industry (most older than 50 years), a good degree of internationalization, and significant R&D expenditure compared with other Italian companies, but still too small to take a strong competitive position in the global pharmaceutical marketplace. A growing body of evidence also highlights the emerging role of SMEs, even among the family firms, that focus on the development of biotechnology drugs or specialize in a particular stage of the innovation process (Farmindustria 2016).

Today, the Italian market for biotech drugs is flourishing; in 2015, it accounted for almost 20% of market share, 40% of new drugs, and 50% of development stage drugs (Farmindustria 2016). In this segment, biotech companies have the choice of specializing in various application fields, including biotech pharmaceuticals, biopharming, vaccines, drug discovery, or drug delivery (i.e., technologies aimed at channeling a pharmaceutical entity to specific areas of the body for specific periods of time) (Farmindustria 2016). It should be noted that though companies are often categorized as biotech when they use biotech techniques as part of their economic activities, pure biotech firms are those that are dedicated exclusively to biotech research activity. Collectively, the biotech sector, at a domestic level, is characterized by a predominance of SMEs with business models that are either *technology-centric* or *know-how-centric* (Farmaindustria 2016). In the former case, firms focus on the development of a broad set of drugs and services based on consolidated technologies, which they manage either in an integrated way (from the upstream stages of research to commercialization) or by granting licenses following basic research and preclinical studies. In the latter case, companies focus on the exploitation of their R&D, regulation, production, or commercialization competencies, by offering them to third parties through strategic partnerships, licensing, and outsourcing agreements. Big companies represent a more limited slice of the market; they are oriented to the final product (*pipeline-centric*) and tend to specialize in the downstream stages of the innovation process that require excellence of skills and resource availability. These enterprises focus on molecules and products that require more time and greater financial investments to develop (Farmaindustria 2016).

To summarize, the Italian pharmaceutical industry structure is an ecosystem of big companies and entrepreneurial SMEs that complement one another and work together in different stages of the innovation process (Arora and Gambardella 1990). This status confirms the importance of a network of highly innovative biotech firms and more traditional pharmaceutical companies as driving forces of industry development and improvements in the quality of cures. Today, innovative processes in the Italian pharmaceutical industry often emerge from a network of different actors; they benefit from the increasing circulation of information and materials among nonprofits, Big Pharma, public research centers, and biotech companies (Farmindustria 2016). In some cases, this collaboration has generated excellent outcomes, such as the first genic therapy, ex vivo, which was the result of strategic cooperation among GlaxoSmithKline, the San Raffaele Telethon Institute for Genic Therapy (Hsr-Tiget), and Molmed, an Italian medical biotechnological company based in Milan (Farmindustria 2016). Therapeutic innovation is therefore the result of strategic agreements among research institutions, large companies, and SMEs that compete with national and international players to exploit research, marketing, operational, and financial synergies. These collaborations are essential, because the ability to achieve profit from innovation also requires a set of complementary assets (manufacturing capabilities, visibility and reputation, marketing capabilities, and regulatory relationships) (Bogner and Thomas 1994; Cool and Schendel 1987; Thomas 1990).

Industry analysis also shows the relevance of institutional settings; the public sector is pervasive with regard to defining the legal and regulatory framework. From the approval process to price definition, public institutions strongly affect industry attractiveness. There is still room for improvement in measures to increase the innovativeness of Italian pharmaceutical companies; they could benefit from (1) developing a regulatory framework to incentivize and standardize innovation and patenting processes; (2) easing the approval process for new drugs by avoiding duplication of efforts to meet fragmented national and regional requirements; (3) favoring access to private equity and venture capital; and (4) increasing the public sector's ability to attract clinical trials to save (public) money and foster innovation.

In the next chapter, a quantitative analysis of technological innovation in the Italian pharmaceutical industry will shed light on the governance structures and innovative performance of Italian pharmaceutical companies. A preliminary examination of the industry confirms that an investigation of the family–innovation relationship in this context is promising for several reasons:

- *Historical relevance.* Today, families control almost 40% of the Italian pharmaceutical industry. Once the owners of local shops, families are now involved in the strategic direction of several Italian companies that aim to preserve values, culture, and a common heritage. For instance, Menarini, Zambon, and Chiesi are currently led by families and feature a strong presence of family board members on their boards of directors (AIDA, 2017).
- *Opportunity from new industry trends.* By challenging the traditional view that the global pharmaceutical industry is an ideal marketplace for large companies, Italian SMEs—many of which are family firms—play a pivotal role in featuring strong research intensity, particularly in innovative technologies and the discovery and development of biopharmaceuticals.
- *Puzzling effect of family firms.* This preliminary overview of the Italian pharmaceutical industry suggests that in this unusual setting, which requires substantial investments in R&D and a strong orientation to innovation, family firms have been able to survive in the domestic competitive arena, in spite of growing international competition and the dramatic transformations caused by continuous technological change. On the one hand, because of the need for relevant investment capacity, family companies may not have the critical mass required to compete effectively, so they should consider equity financing to support innovative products or develop strategic and equity partnerships (both choices that are contrary to the strategic inclination of families to retain full control). On the other hand,

entrepreneurial orientation and emotional attachment can ensure firm survival by feeding the innovation funnel.

Chapter 5 addresses this puzzle by proposing a quantitative investigation of the innovation performance of Italian pharmaceutical firms as a function of a comprehensive set of ownership, management, and governance characteristics.

References

Arora, A., & Gambardella, A. (1990). Complementarity and external linkages: The strategies of the large firms in biotechnology. *The Journal of Industrial Economics, 38*(4), 361–379.

Bierly, P., & Chakrabarti, A. (1996). Generic knowledge strategies in the U.S. pharmaceutical industry. *Strategic Management Journal, 17,* 123–135.

Bogner, W. C., & Thomas, H. (1994). Core competence and competitive advantage: A model and illustrative evidence from the pharmaceutical industry. In G. Hamel & A. Heene (Eds.), *Competence based competition.* Chichester: Wiley.

Bottazzi, G., Dosi, G., Lippi, M., Pammolli, F., & Riccaboni, M. (2001). Innovation and corporate growth in the evolution of the drug industry. *International Journal of Industrial Organization, 19,* 1161–1187.

Bruche, G. (2012). Emerging Indian pharma multinationals: Latecomer catch-up strategies in a globalised high-tech industry. *European Journal of International Management, 6,* 300–322.

Brusoni, S., & Geuna, A. (2003). An international comparison of sectoral knowledge bases: Persistence and integration in the pharmaceutical industry. *Research Policy, 32*(10), 1897–1912.

Brusoni, S., Criscuolo, P., & Geuna, A. (2005). The knowledge bases of the world's largest pharmaceutical groups: what do patent citations to non-patent literature reveal?. *Economics of Innovation and New Technology, 14*(5), 395–415.

Challu, P. (1995). Effects of the monopolistic patenting of medicine in Italy since 1978. *International Journal of Technology Management, 10*(2–3), 237–252.

Chandy, R., Hopstaken, B., Narasimhan, O., & Prabhu, J. (2006). From invention to innovation: Conversion ability in product development. *Journal of Marketing Research, 43,* 494–508.

Cockburn, I. M. (2004). The changing structure of the pharmaceutical industry. *Health Affairs, 23,* 10–22.

Cockburn, I. M. (2006). Is the pharmaceutical industry in a productivity crisis? *Innovation Policy and the Economy, 7,* 1–32.

Cockburn, I. M., & Henderson, R. M. (2000). Publicly funded science and the productivity of the pharmaceutical industry. *Innovation Policy and the Economy, 1,* 1–34.

Cool, K., & Schendel, D. (1987). Strategic group formation and performance: The case of the U.S. pharmaceutical industry. *Management Science, 33,* 1102–1124.

DeCarolis, D. M. (2003). Competencies and imitability in the pharmaceutical industry: An analysis of their relationship with firm performance. *Journal of Management, 29,* 27–50.

DiMasi, J., Hansen, R., Grabowski, H., & Lasagna, L. (1991). Cost of innovation in the pharmaceutical industry. *Journal of Health Economics, 10,* 107–142.

DiMasi, J. A., & Faden, L. B. (2011). Competitiveness in follow-on drug R&D: A race or imitation? *Nature Reviews Drug Discovery, 10*(1), 23–27.

DiMasi, J. A., Hansen, R. W., & Grabowski, H. G. (2003). The price of innovation: New estimates of drug development costs. *Journal of Health Economics, 22*(2), 151–185.

DiMasi, J. A., & Paquette, C. (2004). The economics of follow-on drug research and development: Trends in entry rates and the timing of development. *Pharmacoeconomics, 22,* 1–14.

Faini, R., & Schiantarelli, F. (1983). Regional implications of industrial policy: The Italian case. *Journal of Public Policy, 3*(1), 97–117.

FARMINDUSTRIA. (2016). Rapporto sulle biotecnologie del settore farmaceutico in Italia, available at: www.farmindustria.it.

Fernández, Z., & Nieto, M. J. (2005). Internationalization strategy of small and medium-sized family businesses: Some influential factors. *Family Business Review, 18*(1), 77–89.

Frezza, L. (2015). *Business.* Scienza e Farmaci. L'industria del farmaco: Aboutbooks.

Gamba, S. (2016). *The effect of intellectual property rights on domestic innovation in the pharmaceutical sector*. Bruno Kessler Foundation: Research Institute for the Evaluation of Public Policies.

Gambardella, A. (1995). *Science and innovation: The US pharmaceutical industry in the 1980s*. Cambridge, MA: Cambridge University Press.

Gambardella, A., Orsenigo, L., & Pammolli, F. (2000). Global competitiveness in pharmaceuticals: A European perspective, MPRA Paper No. 15965.

Ghemawat, P. (2010). *Strategy and the business landscape*. Upper Saddle River, NJ: Pearson.

Gómez-Mejía, L. R., Haynes, K. T., Núñez-Nickel, M., Jacobson, K. J., & Moyano-Fuentes, J. (2007). Socioemotional wealth and business risks in family-controlled firms: Evidence from Spanish olive oil mills. *Administrative Science Quarterly, 52*(1), 106–137.

Grabowski, H. G., Kyle, M., Mortimer, R., Long, G., & Kirson, N. (2011). Evolving brand-name and generic drug competition may warrant a revision of the Hatch-Waxman Act. *Health Affairs, 30*(11), 2157–2166.

Grabowski, H. G., & Vernon, J. M. (1987). Pioneers, imitators, and generics—A simulation model of schumpeterian competition. *The Quarterly Journal of Economics, 102*(3), 491–525.

Grewal, R., Chakravarty, A., Ding, M., & Liechty, J. (2008). Counting chickens before the eggs hatch: Associating new product development portfolios with shareholder expectations in the pharmaceutical sector. *International Journal of Research in Marketing, 25*, 261–272.

Griliches, Z., & Cockburn, I. (1994). Generics and new goods in pharmaceutical price indexes. *American Economic Review, 84*(5), 1213–1232.

Henderson, R., & Cockburn, I. (1994). Measuring competence? *Exploring firm effects in pharmaceutical research, Strategic Management Journal, 15*, 63–84.

Henderson, R., & Cockburn, I. (1996). Scale, scope and spillovers: The determinants of research productivity in drug discovery. *Rand Journal of Economics, 27*(1), 32–59.

Hill, C., & Hansen, G. (1991). A longitudinal study of the cause and consequences of changes in diversification in the U.S. pharmaceutical industry 1977–1986. *Strategic Management Journal, 12*, 187–199.

KPMG. (2011). *China's pharmaceutical industry! Poised for the giant leap*. Beijing: KPMG Advisory China Limited.

Levin, R. C., Klevorick, A. K., Nelson, R. R., Winter, S. G., Gilbert, R., & Griliches, Z. (1987). Appropriating the returns from industrial research and development. *Brookings Papers on Economic Activity, 3,* 783–831.

MARKETLINE. (2015). Industrie farmaceutiche italiane, available at: http://www.marketline.com

Orsenigo, L. (1989). *The emergence of biotechnology: institutions and markets in industrial innovation.* London: Pinter Publishers Ltd.

Orsenigo, L., Pammolli, F., & Riccaboni, M. (2001). Technological change and network dynamics: Lessons from the pharmaceutical industry. *Research Policy, 30*(3), 485–508.

Perugini, M. (2015). *Il farsi di una grande impresa.* Franco Angeli Storia: La Montecatini fra le due guerre mondiali.

Petrova, E. (2014). Innovation in the pharmaceutical industry: The process of drug discovery and development. In *Innovation and marketing in the pharmaceutical industry* (pp. 19–81). New York: Springer.

Pisano, G. (1994). Knowledge integration and the locus of learning: An empirical analysis of process development. *Strategic Management Journal, 15,* 85–100.

Powell, W. W., Koput, K. W., & Smith-Doerr, L. (1996). Interorganizational collaboration and the locus of innovation: Networks of learning in biotechnology, *Administrative Science Quarterly, 41*(1), 116–145.

Reiffen, D., & Ward, M. R. (2005). Generic drug industry dynamics. *Review of Economics and Statistics, 87*(1), 37–49.

Roberts, P. W. (1999). Product innovation, product-market competition, and persistent profit- ability in the U.S. pharmaceutical industry. *Strategic Management Journal, 20,* 655–670.

Scalera, V. G., Perri, A., & Mudambi, R. (2015). Managing innovation in emerging economies: Organizational arrangements and resources of foreign MNEs in the Chinese pharmaceutical industry. In *Emerging Economies and Multinational Enterprises* (pp. 201–233). Bingley: Emerald Group Publishing Limited.

Scherer, F., & Weisburst, S. (1995). Economic effects of strengthening pharmaceutical patent protection in Italy. *International Review of Industrial Property and Copyright Law, 26,* 1009–1024.

Sharma, A., & Lacey, N. (2004). Linking product development outcomes to market valuation of the firm: The case of the US pharmaceutical industry. *Journal of Product Innovation Management, 21,* 297–308.

Stremersch, S., & Lemmens, A. (2009). Sales growth of new pharmaceuticals across the globe: The role of regulatory regimes. *Marketing Science, 28*(4), 690–708.

The Economist. (2012). Battling borderless bugs. http://www.economist.com/node/21542410. Accessed 7 Jan 2014.

Thomas, L. G. (1990). Regulation and firm size: FDA impacts on innovation. *Rand Journal of Economics, 21,* 497–517.

Wuyts, S., & Dutta, S. (2008). Licensing exchange—Insights from the biopharmaceutical industry. *International Journal of Research in Marketing, 25*(4), 273–281.

5

Family Business and Technological Innovation: Evidence from the Italian Pharmaceutical Industry

Abstract This chapter analyzes family and non-family firms' innovative activity performance in the Italian pharmaceutical industry. By taking stock of literature on family firms and innovation management, as reviewed in Chaps. 2 and 3, it develops theoretical arguments regarding the multidimensional involvement of families in the innovative performance of Italian pharmaceutical firms, in terms of both the scale and quality of innovation. This chapter includes an empirical section that describes data sources, details the data collection process, and explains the methodology for the data analysis. It concludes by presenting and discussing the empirical findings and drawing conclusions from this evidence.

Keywords Family firms · Innovation performance · Innovation scale and quality · Italian pharmaceutical industry

5.1   Family Firms and Innovation Performance: The Relevance of a Multidimensional Phenomenon

Research on the innovative behavior of family firms has progressed substantially and gained particular momentum in recent years. Increasing recognition of the pervasiveness of family firms across different countries and institutional settings (La Porta et al. 1999; Villalonga and Amit 2009), coupled with growing evidence of the long-term survival and performance of family firms (Family Firm Institute 2017), has spurred interest in how technological innovation occurs in this organizational context (Duran et al. 2016).

Chapter 3 summarized the extant literature on this topic, observing that most empirical research has focused on how different dimensions of family involvement affect both the inputs and outputs of firms' innovative activities. Of the two aspects, inputs have attracted the most scholarly attention, with studies revealing that family firms tend to invest less in innovative activities than non-family firms (Chrisman and Patel 2012; Patel and Chrisman 2014; Anderson et al. 2012; Nieto et al. 2015; Classen et al. 2014; Duran et al. 2016; Broekaert et al. 2016).

The matter of how outputs of innovative activities vary across family and non-family firms is both less investigated and more ambiguous. Theoretically, it is important to recognize that the drivers of innovation inputs and outputs are different. Innovation inputs are mainly the result of deliberate managerial decisions, but innovation outputs depend on more than just the financial resources committed to innovation activities. Rather, they are influenced positively or negatively by firms' technological postures and competencies, i.e., the set of resources and capabilities (e.g., specialized human resources, knowledge, routines, technological capabilities) that can be mobilized for innovation objectives (Matzler et al. 2015). Therefore, though it is vitally important to understand the family firm approach to investments in innovation inputs, an analysis of family firms' innovation outputs may offer an even more meaningful explanation of the specific nature of innovative activities in such firms.

There are challenges to empirically capturing the performance of innovative activities though, because innovation outcomes can take highly heterogeneous forms. In some contexts, the number of patents produced may be a good proxy of innovative performance (Levin et al. 1987), but in other settings, upstream innovative activities are less relevant; the number of new product introductions may offer a more appropriate indicator of firms' innovative capabilities. For these reasons, current understanding of the innovative performance of family firms is limited, and additional research is needed.

This chapter contributes to this area of investigation by focusing on the multidimensionality that characterizes the constructs of family influence and innovation performance. Citing previous research (e.g., Block 2012; Block et al. 2013; Matzler et al. 2015), Chaps. 2 and 3 emphasized that the heterogeneity of family firms stems from the family's involvement in various aspects of firm ownership, management, and governance. Previous theoretical and empirical research in the area of family business has demonstrated that the nature of firms' innovative behaviors depends on whether the firms are merely owned, or are owned and managed, by family members; it is thus important to distinguish between ownership and management dimensions (Block 2012; Block et al. 2013; Matzler et al. 2015). Theoretical insights and empirical findings, though, are not univocal. For example, researchers have argued that family managers pursue the survival of their firms, in line with the interests of family owners but to the detriment of firm innovation (Block et al. 2013). However, other authors show that family managers are less concerned with short-term performance and often make strategic decisions based on long-term scenarios, thus promoting valuable innovation projects (Matzler et al. 2015). Similarly, though family-influenced boards of directors have been associated with constraining, parsimonious approaches to innovative activity (Banno 2016), it has also been suggested that they support enriched uses of innovation inputs (Matzler et al. 2015). Thus, it is critical to disentangle how the individual dimensions of family firm involvement affect the technological performance of family firms.

This chapter proposes that in addition to accounting for the multidimensionality of family involvement, scholars of family business should recognize that also innovation performance is a complex, multifaceted phenomenon (Vanhaverbeke et al. 2014) that requires research beyond studies of single aspects of firms' innovative performance (Lodh et al. 2014; Kraiczy et al. 2014, 2015; Banno 2016).

Empirical research shows great variation in the different qualitative dimensions of innovations (e.g., Trajtenberg 1990; Trajtenberg et al. 1997; Hall et al. 2001; Gambardella et al. 2008), documenting that all innovations are not created equal. In many industries, there is a clear division of labor in terms of the nature of innovative activity (Gambardella 2005; Lee and Berente 2012). For example, within the same sectors, some firms focus on scientific discovery and advancement of the technological frontier, while others invest in applied, downstream innovative activities (Gambardella 2005). Because firms that compete in the same innovative ecosystem may seek to distinguish themselves by privileging specific aspects of the broader innovation process, it may be misleading to observe only one aspect of a firm's innovative performance. When firms choose to devote their innovative efforts to the development of specific abilities, their performance along other dimensions of the innovative activity is affected. In the context of family firms, the specificities of the family's objectives, resources, and incentives may favor distinct technological competencies that lead to better performance in specific areas of innovation. Thus, innovative performance can be understood only by recognizing its multidimensionality (Vanhaverbeke et al. 2014).

This chapter investigates the innovation performance of family firms by accounting for both quantitative and qualitative dimensions of this construct (Phene and Almeida 2008), and by unpacking the qualitative dimension to account for the complex nature of innovation in the pharmaceutical industry. In this setting, innovation and new knowledge creation are critical to firm survival, but firms may choose to specialize in very different innovative tasks. As Chap. 4 shows, actors in this industry range from small, entrepreneurial biotech firms that push the technological frontier through leading-edge discoveries in fundamental science, to large, diversified pharmaceutical multinationals that engage in all stages of the innovation process. Although both types of firms can be successful innovators, they are likely to substantially differ in terms of their innovative capabilities and, presumably, will excel in very different areas of technological competence.

5.2 Scale and Quality of Innovation in Family Firms: A Tale of Two Relationships?

Several theoretical perspectives offer insights to understand the relationship between family influence and innovation performance.

The resource-based view suggests that family firms possess unique knowledge assets, relational resources, and competencies that make highly productive the use of (often limited) innovation inputs (Matzler et al. 2015; Duran et al. 2016), thereby envisaging a positive effect of family influence on firm innovation performance.

Conventional agency theory suggests that the ownership and management of family firms are strongly aligned, because managers belong to the families that own the companies (Jensen and Meckling 1976). In this condition, agency costs decrease, and more virtuous innovation behavior is encouraged (Munari et al. 2010).

However, this view has been challenged by both theoretical and empirical insights. For example, research suggests that family owners tend to be more risk-averse than their non-family counterparts, because their wealth is likely to be more concentrated in their companies (Basu et al. 2009; McConaughy et al. 2001; Mishra and McConaughy 1999). This scenario may drive family members to avoid making decisions that increase their firms' overall business risks, thus lowering their involvement in innovation projects and, in turn, their firms' innovative performance.

Advancing traditional agency perspectives, the behavioral agency model (BAM) has been originally proposed to challenge established assumptions regarding executives' risk profiles (Wiseman and Gómez-Mejía 1998). Building upon this model, family business scholars argue that family owners are averse to the loss of socio-emotional wealth (SEW) (Gomez-Mejia et al. 2007) that tends to concentrate in family firms. Thus, family owners tend to behave in ways that ensure continuity and control, even when such behavior collides with economic goals. To avoid the loss of SEW, members of family firms may be willing to bear significant risks, but at the same time, could shy away from hazardous business decisions that might endanger the status quo. In other words, according to the BAM, family firms may be simultaneously risk-inclined and risk-averse (Gomez-Mejia et al. 2007). This potentially divergent attitude of the family toward the firm's business risk suggests that predicting whether the relationship between family influence and innovation performance will be positive or negative is unlikely to be an easy task.

To envisage how this relationship looks like in a BAM perspective, it is important to remember that the innovation performance is not only determined by (1) the firm's R&D investment but also influenced by (2) the set of resources, capabilities, and processes the firm can mobilize for innovation purposes, as well as by (3) the firm's innovation orientation and technological posture. Therefore, there are at least three channels of family influence on firm innovation output that should be considered. First, in terms of R&D investment, research has found that willingness to preserve family security drives family firms to underinvest in risky R&D activities that may divert resources from existing businesses and expose the firm's SEW to substantial and undesired changes (Chrisman and Patel 2012; Anderson et al. 2012; Chen and Hsu 2009; Patel and Chrisman 2014). Second, in terms of the resources and capabilities available for innovative activities, a family loss aversion may lead to nepotistic behaviors (Sirmon and Hitt 2003), such as hiring family members regardless of their technical competencies, adoption of questionable promotion practices, or implementation of economically irrational systems of incentives (Le Breton-Miller et al. 2011; Schulze et al. 2001). These behaviors likely hinder the process through which firms transform innovation inputs into innovation outputs. Third, in terms of firm innovation orientation, when the amount of resources available for the development of innovative activities is kept constant, the type of innovative outputs that a firm achieves may vary significantly, depending on the firm's inherent strategic goals and technological posture. In this respect, family firms' typical concerns for continuity and SEW preservation may lead to adopt a very incremental—as opposed to a more radical—innovation orientation, to the aim of stabilizing the firm's current business and strategy (Carnes and Ireland 2013). Such an incremental, exploitative approach to technological innovation is likely to constrain the family firm's ability to contribute significantly to technological progress.

Overall, these insights suggest that, even within a BAM perspective, family-owned firms are likely to underperform, in terms of innovation outputs, compared with non-family firms. The non-pecuniary goals of family firms drive them to lower their strategic commitment to innovative activities and adopt nepotistic practices that can have detrimental effects on innovation.

These predictions, which usually are conceived in relation to family ownership, may be valid also in relation to both the management and governance dimensions of the family. On the one hand, because the interests of family managers are aligned with those of family owners, and because such managers can more actively and directly shape their firms' strategies, they have the power to reduce involvement in innovative activities that threaten the firms' stability and endurance (Block et al. 2013). On the other hand, family-dependent boards of directors can advise against the pursuit of innovative projects that expose their firms to higher risks and noneconomic losses; they provide additional protection to family interests by supporting and counseling family managers as they pursue the ultimate objective of perpetuating SEW (Corbetta and Salvato 2004).

The prediction of an overall negative influence of the family on firm innovation output, however, should be refined to account for the specificities of this study's context and to acknowledge the potentially heterogeneous effects on the distinct dimensions of innovation output.

The technology-intense and innovation-critical nature of the pharmaceutical industry offers a particularly interesting setting for studying the role of family involvement in firms' technological innovation. Firms that are unable to innovate, or that deliberately avoid innovation and the development of new technological knowledge, may find it difficult to confront the challenges of turbulent, dynamic, high-technology environments; they risk being pushed out of the competitive arena (Teece et al. 1997). By staying actively engaged in innovation, firms can stay abreast of and adapt to key technological disruptions, thus increasing their chances of long-term survival. In other words, in high-technology industries, it is the avoidance of innovation, rather than the continuous development of new technological knowledge, that carries the greatest risk; resistance to change incurs the greatest losses, because it prevents firms from dynamically adapting to the evolution of the external environment. Thus, contrary to theoretical expectations, in high-technology environments family firms may be willing to generate a similar or even greater scale of innovation than non-family firms. It may be the case that family members realize that the specificities of the business contexts in which they operate make innovation essential to mitigating, rather than increasing, their level of business risk (Gomez-Mejia et al. 2014). Therefore, contrary to established assumptions about goal conflict in family firms, the conservative tendencies of family business owners and the economic objectives of their firms do not necessarily collide (Chrisman and Patel 2012). Rather, goal congruence may occur when family firms' long-term objectives of continuity, along with the technological intensity of the industry context, orient them toward innovation to ensure firm survival.

Very different considerations may apply to the qualitative dimensions of innovation, allowing better definition and understanding of the intrinsic nature of firms' innovation strategies. Consistent with their orientation to the long term (Sirmon and Hitt 2003), family firms may rationally choose to engage in a significant amount of innovation to adapt to rapidly changing technological settings, with the ultimate objective of ensuring firm longevity. However, the nature and content of the innovations they develop may differ according to their risk tolerance, resources, and incentives. Hence, even in contexts in which innovation is an inevitable component of firm strategy, once firm survival is secured through the creation of a substantial scale of innovations, family firm's inherent risk aversion could prevail: such aversion could drive them to select mainly incremental, exploitative innovation projects that have limited potential to generate significant breakthroughs. These projects allow risk-averse family firms to reduce their involvement in highly uncertain endeavors, thus minimizing the chances of catastrophic consequences associated with project failure. Moreover, the ability to perform well on specific qualitative dimensions of technological innovation is more likely than the ability to generate innovations to be affected by the firm's technological posture and the capabilities, resources, and processes it can leverage to support its innovative activities. For example, the adoption of nepotistic behaviors can severely inhibit a firm's ability to spawn innovative outputs that are at the forefront of the technological frontier. When family members or friends are appointed to key managerial and technical positions, irrespective of their actual education and skills (Le Breton-Miller et al. 2011; Schulze et al. 2001), the conditions that allow for the generation of highly novel, fundamental knowledge are likely to be jeopardized; harm is caused by both a lack of appropriately trained human resources and a reduction in incentives for qualified employees, who may underperform because of unfair hiring and promotion practices. Therefore, even when there is a significant commitment of financial resources to innovative activities, a lack of professional managers and knowledge workers, combined with the lack of a clear orientation to specific and ambitious technological objectives, could constrain the ability of family firms to excel along the distinct dimensions of innovation quality.

5.3 Data and Empirical Methodology

The objective of this study is to analyze empirically how different dimensions of family involvement in the firm's business influence the innovative performance of Italian companies in technology-intense sectors, specifically the pharmaceutical industry. Data availability (discussed in greater detail in Chap. 6) makes this undertaking particularly challenging. To pursue its objective, this study leverages a blended approach that uses a quantitative analysis of a sample of Italian pharmaceutical companies, combined with qualitative evidence gathered from face-to-face interviews with owners and managers of a subset of sample firms and other industry experts[1] (for a similar approach, see Scalera et al. 2014).

The empirical strategy consists of a series of steps aimed at identifying a representative sample of Italian pharmaceutical companies. The resulting sample enables the isolation of the effect of family on various aspects of firm innovation. The sample of Italian firms is generated from the Italian Digital Database of Companies (AIDA) produced by Bureau van Dijk. The AIDA includes the financial and governance information of all public and private firms in Italy, with data collected from official sources and reported annually to the Italian Chamber of Commerce. Thus, firms operating in the *manufacture of basic pharmaceutical products and pharmaceutical preparations* (code NACE 21) were identified, resulting in a sample of 709 firms. To focus on the Italian industrial system, the sample was narrowed to 320 companies owned by at least one shareholder located in Italy (IT), with 51–100% of firm equity. From this list, the final sample of the top 50 firms, ranked by revenue, was determined. This sample is highly representative of the entire Italian pharmaceutical industry since it accounts for almost 90% of total revenues and 84% of total employees from the original sample of 329 firms.

To analyze the innovative performance of Italian pharmaceutical firms, this study relies on information available in patent documents. Patents are a primary data source for analyzing the innovative profile of pharmaceutical companies. The wide use of patents in the industry to protect the outcomes of firms' innovative efforts (Levin et al. 1987; Bierly and Chakrabarti 1996) justifies the use of patent data to analyse firms' innovative outputs. The patent portfolios of the sample firms analyzed in this study have been reconstructed using data published by the US Patent and Trademark Office (USPTO). The choice to rely on US patents is justified by the features of the empirical context. The pharmaceutical industry is increasingly regarded as a global industry; this condition compels companies affiliated with the sector to protect the outcomes of their innovative activities in every relevant national market. The USA is still the primary market for pharmaceutical products, so pharmaceutical companies, regardless of their home countries or the geographical locations in which they conduct innovative activity, likely apply for patent protection in the USA. They are even more likely to do so, considering the overall costs associated with the pharmaceutical innovation process. Although firms in many other industries might be unwilling to apply for patent protection in a foreign country such as the USA because of associated costs, an informant interviewed during the research process confirmed that patent-seeking costs represent a trivial component of the average pharmaceutical firm's overall investment in innovation, such that firms commonly expand their request for intellectual property protection in every country in which they have interests, and especially in the USA.

Patent documents are an extremely rich source of innovation-activity information. Each patent document pinpoints the organization to which the patent's ownership has been assigned, the technological classes with which the invention is associated, and the temporal characteristics of the invention (i.e., dates at which the patent was applied for and granted) (Trajtenberg 1990). Patent documents also identify *forward citations*, i.e., the follow-on inventions that cite the patent as prior art (Jaffe et al. 1993) and the scientific literature used as knowledge sources during the invention process. To facilitate the identification of patents assigned to the sample firms, this study also relies on the online data set published by Martin Goossen, which offers access to clean data for all USPTO patents granted between 1976 and 2016.[2] All USPTO patents assigned to any of the companies included in the original sample were selected on the basis of the assignee-name field.

The independent variables considered in this study seek to identify the different roles of the family in firms operating in the Italian pharmaceutical industry. To account for family firm heterogeneity, as described in Chap. 1, hand-collected data differentiate the dimensions of family business—i.e., family ownership, governance, and management—for the period from 2006 to 2013. Governance and management dimensions of the sample firms are reconstructed by collecting all available information from public official filings deposited in the Italian Chamber of Commerce. These official filings are structured in the form of historical reports that contain all changes that have occurred in the firms' governance structure and characteristics during 2006–2013. The information includes: (1) the main characteristics of each shareholder, (2) leadership team of the firm, and (3) detailed information about the board of directors. Following the methodology of Miller et al. (2013), the data reveal both family-related board members and CEOs by their surname affinity with that of the major shareholder. Finally, data on ownership structure and characteristics obtained from AIDA were double-checked with ownership information collected from public official filings provided to the Italian Chamber of Commerce.

The empirical analysis covers the period from 2006 to 2013; the first year of the observation window (2006) is determined by the availability of data on firm ownership, governance, and management, and the last year (2013) ensures an appropriate collection of patent data, which could be affected by right-truncation biases. The patent data pertain to granted patents; in general, there is a lag of about two years before patents are granted (Bloom and Van Reenen 2002).

5.3.1 Variables

Dependent Variables

Previous research has documented that innovations vary substantially along different qualitative traits (e.g., Trajtenberg 1990; Trajtenberg et al. 1997; Hall et al. 2001; Gambardella et al.2008). This suggests that innovation performance is a multidimensional, complex construct (Vanhaverbeke et al. 2014). As a consequence, it can be understood only by accounting for several facets of this general concept. To this aim, this study unpacks innovation performance by considering both its quantitative and qualitative dimensions (Phene and Almeida 2008).

In the previous literature on the pharmaceutical industry, innovation performance has been investigated by analyzing a firm's patent portfolio (Gambardella 1992; Achilladelis and Antonakis 2001; Cockburn and Henderson 2000; De Carolis 2003; Nerkar and Paruchuri 2005; Hess and Rothaermel 2011). As the foregoing discussion explains, patents are used extensively in this industry because they provide significant protection (Levin et al. 1987). Therefore, by reconstructing a firm's patent portfolio, it is possible to trace the outcome of its innovative activities. Moreover, the use of patent information enables analyses of both the quantitative and qualitative dimensions of a firm's innovation performance. Although the number of patents a firm files is clearly an indicator of its scale of innovation, more in-depth analyses of patent characteristics also allow for evaluations of several qualities of the underlying inventions.

With this rationale, this study focuses on patent production as a measure of the scale of innovation. It captures the quality of a firm's innovation by analyzing other dimensions of patent production derived from previous literature, i.e., innovation value, basicness, and technological scope. Each of these variables is explained next.

Innovation scale. The *scale* of a firm's innovation is measured by counting the successful patent applications a firm has filed in a given year *t*. The choice of using the year of patent application rather than the year of patent grant is deliberate and justified by the fact that the patent application date is closer to the time when the invention underlying the patent was developed, compared with the date of the patent grant. The patent grant date likely depends on the process of patent examination, which could differ across patents for several reasons. Previous research shows that for companies in a high-technology industry such as pharmaceuticals, the ability to file for significant numbers of patents is a critical measure of upstream innovative performance (Penner-Hahn and Shaver 2005).

Innovation basicness. In the pharmaceutical industry, reliance on science is often critical for successful innovation, because basic knowledge can spur new technology developments. Previous research suggests that the basicness of inventions can be measured by the number of times the patents associated to such inventions cite scientific publications as prior art. Scientific citations capture the knowledge flows between science and technology (Tijssen 2001). They signal the innovative firm's ability to decode and make productive use of the advances that occur in the scientific world of fundamental knowledge (Cassiman et al.2008). More generally, scientific citations allow the qualification of scientific content of patents by measuring patents' proximities to the scientific frontier of knowledge activities (OECD 2011). Specifically, this measure is built by summing the scientific citations referenced by each patent that a firm successfully applied for in year t.

Innovation value. Although the number of patent applications a firm successfully files provides information on the scale of its innovation, it does not reveal anything about the value of the patents. Literature suggests that most patent applications filed by companies are never used for market commercialization, which suggests their value is limited (Chesbrough 2006). To assess the value of a firm's patents, established research in technology and innovation management uses forward citations (Trajtenberg 1990; Hall and Ziedonis 2001). Forward citations describe the links between a patent and the set of follow-on inventions that build on the patent's underlying innovation. Frequent citation of a patent shows that it has been used to develop subsequent technical advances and can thus be considered of high value. Previous research suggests that forward citations capture the economic value of a firm's body of innovation (Hall and Ziedonis 2001) and are correlated with the firm's market value (Hall et al. 2005). When building this variable, it is important to recognize that forward citations are accumulated over time as applications for subsequent patents appear that employ the related innovation as a knowledge source. Therefore, the number of forward citations a patent receives is likely to depend on its age, because older patents have been available to be cited for a longer period of time. To address this potential truncation bias, the models that estimate *innovation value* include a control for *patent age* (Gittelman and Kogut 2003). This measure thus is built by summing the forward citations for each patent that a firm successfully applied for in year t, until 2016, and adding a control for the number of years these patents have been available to be cited (i.e., 2016 minus the patent application year) to the regressions.

Technological scope. Patent documents report one or more technological classes that identify the technological fields to which a patent pertains, assigned to the patent through an accurate process. Lerner (1994) suggests that the number of different technological classes to which a patent is assigned defines its *technological scope*. In other words, it captures the extent to which a patent is relevant to different technological fields. This study adapts this approach to evaluate the technological scope of the firm as a whole (rather than of its individual patents) by analyzing the number of different international four-digit patent classes (IPC) associated with the whole set of the firm's patents applied for in year t. A firm that is able to develop inventions in different technological fields is likely to possess a wide technology base; therefore, the higher the number of different technological classes in which a firm has filed patents in year t, the broader the scope of its technological capabilities.

Independent Variables

Despite the diffusion of family firms around the world, researchers disagree on a clear definition of the business type. However, there are three dimensions that are commonly used to describe family firms and capture their heterogeneity; they are essential for determining a family's influence on the firm's strategic decision (Villalonga and Amit 2006; Schmid et al.2015). First, *family ownership* identifies firm shareholders. Second, *family management* focuses on the firm's CEO to understand who manages the company and has the power and authority to undertake or supervise strategic decisions about the firm's innovative activities. Third, *family governance and control* investigates the composition of the board of directors as the entity that monitors and controls the company, appoints the CEO and top management team, and advises organizational decision makers (for a review of the role of the board in family business, see Bammens et al. 2011). This study adopts these distinctions to disentangle the heterogeneous nature of family business.

Family ownership is the total number of shares owned by members of the dominating family/ies (and their relatives) divided by the total shares outstanding (e.g., Anderson et al. 2012; Matzler et al. 2015). To identify shares held by dominating families and their relatives, the criterion of surname affinity applies; all stocks owned by the different family owners are summed (Miller et al. 2013). Data on ownership structure are drawn from AIDA and double-checked with the official filings with the Italian Chamber of Commerce.

Family management is measured by two variables. The first captures the presence of a *family CEO* through a dummy that is equal to 1 if the CEO belongs to the dominating family and 0 otherwise (Miller et al. 2013; Block 2012). Surname affinity with the dominating family also informs this variable (Miller et al. 2013). The second variable explores the leadership structure of the firm, accounting for the effect of *CEO duality*. This variable is equal to 1 if the CEO is also the chair of the board and 0 otherwise. Data on family management variables are manually coded, according to information included in the official filings with the Italian Chamber of Commerce.

Family governance is captured by *family supervision*, a variable that is calculated as the ratio of the directors belonging to the dominating family to the total number of directors in the board, as a percentage (Schmid et al.2014). Data on family supervision come from official filings with the Italian Chamber of Commerce.

Control Variables

The empirical analysis controls for factors that could affect innovation performance, as identified by prior research. Following previous studies that use innovation performance as the dependent variable, selected control variables lagged at $t-1$ serve to address the reverse causality issue derived from the potential simultaneity of innovation performance and some of the control variables included in the model.

ROA. Firm performance is measured by the return on assets (*ROA*) in year $t-1$, to account for firm efficiency (e.g., Matzler et al. 2015). Scholars of family business have emphasized the influence of prior performance on firm innovation performance. In particular, when prior firm performance is below aspirations, family firms can decide to prioritize financial over non-financial goals, which positively affects firms' innovation posture (Chrisman and Patel 2012).

Size. The empirical analysis also controls for firm *size*, as captured by the firm's number of employees in year $t-1$ (Spriggs et al. 2013; Kraiczy et al. 2015; Nieto et al. 2015). There is no established agreement on the impact of firm size on innovation. On the one hand, greater size may generate scale economies in R&D or increase the availability of internal resources that could be devoted to innovative activities. On the other hand, larger firms offer fewer incentives to individual scientists, are more sensitive to losses in managerial control, and tend to be less flexible and more bureaucratic than smaller firms, with obvious effects on the firm's innovative outcomes.

Patent stock/patents. In models that estimate the *scale* of firm innovation, a variable that captures the (logarithm) of the numbers of patents the firm has applied for in the four years prior to year t (plus one) is included to control for the firm's prior technological capability (Wadhwa and Kotha 2006). In models that estimate *innovation value, basicness,* and *technological scope,* a variable equal to the (logarithm) of the number of patents the firm applied for in year t (plus one) is included in the regressions, because the three dependent variables used in these models should vary with the number of patents applied for in year t. These two controls are not included simultaneously in the models, because they tend to be highly correlated.

Collaboration. The performance of a firm's innovative activities may be affected by the extent to which it is open to collaboration with external partners. To account for this effect, a dummy variable takes the value of 1 if the firm has successfully filed for at least one coassigned patent in the year $t-1$. Coassigned patents are relatively frequent in the pharmaceutical industry (Hagedoorn 2003); they represent the outcome of joint innovative activities that may contribute to the firm's innovation funnel (Perri et al. 2017).

Experience. The firm's innovative experience is measured as the number of years between the firm's first successful patent application and year t. This variable is particularly relevant in the context of this analysis, because it partially captures the distinction between more established companies and younger, more entrepreneurial organizations such as biotech firms.

Technological diversity. The analysis also accounts for differences in the technological diversity of firms' innovation pipelines. Firms may be more or less focused in the technological paths they pursue in any given period of time; these paths may influence their prospective innovative endeavors. For each firm i, the index is expressed as:

$$Technological\ diversity_i = 1 - \sum_{j=1}^{J}(s_{ij})^2,$$

where s_{ij} is the percentage of the successful patents applied for by the firm i in year $t-1$ that belong to the technology class j. The index could vary between 0 and 1. It scores 0 for firms whose patents concentrate in the same technological class, and it asymptotically approaches 1 when the firm's patents are technologically extremely dispersed. Some of the firms in our sample have no patents in specific years. When this happens, this variable is undefined. To avoid dropping these observations, we set this variable to 1 as we assume that, in these years, the firm does not have any specific technological focus to shape its future innovative activity.

5.3.2 Data Analysis

The data structure (multiple observations for firms at different points in time) requires a panel data analysis. Specifically, the data set is organized as an unbalanced panel. The patent-based dependent variables are all count variables that take nonnegative integer values (e.g., Almeida and Phene 2004; Phene and Almeida 2008; Block et al. 2013). To deal with such dependent variables, a Poisson model often is used, but patent-based dependent variables tend to be characterized by the presence of overdispersion, a condition that arises when the standard deviation exceeds the mean. This condition clearly occurs in this study sample, due to the large variance that characterizes not only the *innovation scale* variable (standard deviation/mean = 3.14, see Table 5.1) but also the qualitative innovation performance variables: *innovation basicness* (standard deviation/mean = 4.23, Table 5.1), *innovation value* (standard deviation/mean = 3.96, Table 5.1), and *technological scope* (standard deviation/mean = 3.33, Table 5.1). In the presence of overdispersion, the Poisson model might underestimate the standard errors and inflate significance levels. Therefore, following previous research (Almeida and Phene 2004; Phene and Almeida 2008; Perri and Andersson 2014), this study uses negative binomial model regression, an econometric technique that corrects for overdispersion.

To avoid potential problems associated with simultaneous causality and to facilitate causal inference, a one-year lag between the dependent

Table 5.1 Mean, standard deviation, minimum, and maximum values

	Observations	Mean	Standard deviation	Min	Max
Innovation scale	400	1.70	5.33	0.00	41.00
Innovation basicness	400	16.30	68.88	0.00	977.00
Innovation value	400	1.33	5.27	0.00	41.00
Technological scope	400	2.92	9.71	0.00	86.00
Family ownership	347	84.06	25.04	0.00	100.00
Family supervision	382	0.48	0.29	0.00	1.00
CEO duality	382	0.27	0.45	0.00	1.00
Family CEO	382	0.57	0.50	0.00	1.00
Patent stock	400	0.87	1.28	0.00	5.04
ROA	389	7.86	9.23	−43.14	42.81
Size	381	320.90	358.13	3.00	1643.00
Collaboration	400	0.02	0.13	0	1
Experience	400	11.85	12.90	0	41
Technological diversity	400	0.86	0.28	0	1

and independent variables is applied for all models reported (e.g., Munari et al. 2010).

5.4 Analyses and Results

5.4.1 Descriptive Statistics

The descriptive statistics are in Tables 5.1, 5.2, and 5.3. Table 5.1 shows the means, standard deviations, and the minimum and maximum values for the innovation variables, governance variables, and control variables used in the analysis. Table 5.2 presents the percentiles of governance (*family ownership, family CEO, CEO duality, family supervision*) and innovation (*innovation scale, basicness, innovation value*, and *technological scope*) variables, and Table 5.3 reports the correlation matrix.

In the first table, the means, medians, and standard deviations of the variables for all firm-years are shown. As previously mentioned, the sample includes the 50 largest pharmaceutical Italian firms as of 2016 and spans 2006–2013, yielding 400 firm-year observations.

With regard to the innovation variables and specifically *innovation scale*, the data show that firms in the sample apply for an average of 1.70 *patents* per year. Although the standard deviation and the maximum value of the number of *patents* are, respectively, 5.33 and 41.00—suggesting that some of the sampled companies generate a significant number of inventions—these figures overall imply very limited patent production, particularly given the technological intensity of the industry being analyzed. With regard to the qualitative dimensions of innovation (*innovation basicness, innovation value, and technological scope*), the values observed should be assessed according to data on the number of patents, because they may be influenced by the raw number of patent applications filed by the sample firms. Starting with *innovation basicness*, Table 5.1 reveals that in line with the profile of the industry—which relies on basic knowledge and scientific advances as inputs to the innovation process—firms in the sample make an average of 16.30 scientific citations per year, though this number is subject to significant variation, as highlighted by a standard deviation of 68.88 and a maximum

Table 5.2 Percentiles of governance and innovation variables

Percentiles of governance variables				
Percentiles (%)	Family ownership	Family supervision	Family CEO	CEO duality
5	28.31	0.00	0.00	0.00
10	42.50	0.14	0.00	0.00
25	72.00	0.25	0.00	0.00
50	100.00	0.43	1.00	1.00
75	100.00	0.67	1.00	1.00
95	100.00	1.00	1.00	1.00

Percentiles of innovation variables				
Percentiles (%)	Innovation scale	Innovation basicness	Innovation value	Technological scope
10	0.00	0.00	0.00	0.00
25	0.00	0.00	0.00	0.00
50	0.00	0.00	0.00	0.00
75	1.00	2.00	0.00	1.50
90	3.00	38.00	1.00	6.00
95	10.50	76.00	11.00	18.50
99	37.00	282.50	32.00	58.50

value of 977.00 citations per year. With regard to *innovation value*, the number of forward citations of firms' patents has a mean value of 1.33, which suggests that patent portfolio values, on average, are relatively low, despite a standard deviation of 5.27 and a maximum value 41.00. Finally, with regard to *technological scope*, the average number of technological classes is 2.92, which suggests that firms in the sample are able to span an average of three different technological fields with their yearly inventions. The standard deviation of 9.71 and the maximum value of 86.00 point to significant variation in firms' abilities to broaden their technological scope though. As reported in Table 5.2, the medians (50th percentile) of these variables are all 0, consistent with prior work on innovation in family businesses (e.g., Hsu et al. 2014).

The summary statistics for governance variables show that the average fraction of shares in the hands of the family is 84.06%, corresponding to a median value of 100%. On average, 86% of sample firms own the majority of ownership stakes (>50%). It is interesting to note that though firms' ownership structures reveal a high-ownership concentration, families open their boards of directors to non-family members. On

Table 5.3 Correlation matrix

	1	2	3	4	5	6	7	8	9	10	11	12	13
(1) Innovation scale	1.00												
(2) Innovation basicness	0.54*	1.00											
(3) Innovation value	0.71*	0.51*	1.00										
(4) Technological scope	0.98*	0.51*	0.67*	1.00									
(5) Family supervision	0.08*	-0.01	0.07	0.08	1.00								
(6) Family CEO	0.03	0.08	0.10	0.03	0.56*	1.00							
(7) CEO duality	0.01	0.15*	0.09	0.00	0.25*	0.42*	1.00						
(8) Family ownership	0.16*	0.12*	0.10	0.15	-0.19*	-0.22*	0.11*	1.00					
(9) Patent stock	0.74*	0.51*	0.56*	0.71*	-0.06	-0.03	0.06	0.24*	1.00				
(10) ROA	-0.17*	-0.08	-0.07	-0.16*	-0.02	0.01	-0.03	-0.06	-0.22*	1.00			
(11) Size	0.43*	0.26*	0.32*	0.42*	-0.03	-0.03	0.03	0.11	0.64*	-0.01	1.00		
(12) Collaboration	0.38*	0.61*	0.54*	0.32*	-0.01	0.04	0.09	0.05	0.33*	-0.08	0.11*	1.00	
(13) Experience	0.44*	0.32*	0.32*	0.42*	-0.14*	-0.11*	0.03	0.16	0.69*	-0.20*	0.61*	0.20	1.00
(14) Technological diversity	-0.16*	-0.14*	-0.13*	-0.13*	0.06	0.05	-0.01	-0.12	-0.53*	0.16*	-0.35*	-0.03	-0.38

$*p < 0.05$

average, family members hold 48% of all director seats, with a standard deviation of 0.29. Also surprisingly, Table 5.3 reports a slightly negative and significant correlation between *family ownership* and *family supervision* (−0.19 at $p < 0.05$). However, for 51% of the firms, nearly half of director seats belong (on average) to family members, showing not only the relevance of family firms in monitoring and advising top management's decision but also their need to involve non-family board members. This practice is becoming common in Italian pharmaceutical companies. In a face-to-face interview conducted during the data collection process, the president of an Italian pharmaceutical company, explaining the logic underlying board composition, explicitly noted the relevance of appointing non-family board members, highlighting that two new non-family directors had been appointed to his firm's board. The first was an experienced manager who had worked for a long time in the company and had extensive knowledge of the pharmaceutical industry; the second was an independent director with no prior connections to the company. This professional board member worked for a pharmaceutical company in the past and now leads a multinational company in a different industry. The company informant believes that the new independent director will contribute a fresh view and strong international experience to the firm, because he previously worked for a multinational company controlled by a family; there is an underlying expectation that his lower emotional attachment to the firm can help manage potential conflicts between family members and the firm.

However, family firms also can use different channels to influence firm actions. The possibility of exerting an influence on firm policy is stronger when family members have an active management role. It is not surprising that, during 2006–2013, only 34% of firms in the sample did not experience the direct involvement of family members in the CEO position. An analysis of *family CEO* by year reveals that over time, Italian pharmaceutical firms have appointed an increasing number of family CEOs. The average value for the *family CEO* variable is 0.57, and the median value is 1.00, confirming the high propensity of family members to lead pharmaceutical companies in Italy. The governance variable of *CEO duality* investigates firm leadership structure (e.g., Krause 2017). During 2006–2013, 44% of the sample firms had combined CEO and board chair positions for at least one year. Evidence from the leadership structure of Italian pharmaceutical firms reveals that in 24% of firm-years observations, the family CEO also holds the board chair position, exhibiting strong unity of command in the hands of family members. These statistics are noteworthy because they provide preliminary evidence that family members, though not often involved in the administration of their businesses, tend to use their ownership positions to monitor and advise management.

Table 5.3 displays a correlation matrix for all variables in the sample. As expected (e.g., Block et al. 2013), the correlation between *family supervision* with *family CEO* is positive and significant (r = 0.56, p < 0.05). Results also display other potential sources of multicollinearity which, as explained below, have been checked to ensure robustness to the multivariate analysis, i.e., a positive and significant correlation between firm *size* and *experience*, and between *firm size* and *patent stock* (0.61 and 0.64, respectively), a negative a significant correlation between *technological diversity* and *patent stock* (−0.53), and a positive and significant correlation between firm *patent stock* and *experience* (0.61).

5.4.2 Univariate Analysis

Table 5.4 reports the differences of means tests, based on governance variables. The performance of firms' technological innovation and control variables are analyzed to account for the heterogeneity of family firms in term of ownership, management, and governance. In Panel A of Table 5.4, the differences in means among firms with low (below the mean) and high (above the mean) *family ownership* are reported. Starting with the *innovation scale*, the first subsample shows a significantly lower number of *patents* (1.05 versus 2.17, p = 0.019); findings on the quality of innovation are mixed. In terms of the degree of *family ownership*, low- and high-ownership firms exhibit no significant differences in *innovation value*, but high-ownership firms register significantly higher *basicness* (21.88 versus 8.46, p = 0.027) and *technological scope* (3.73 versus 1.77, p = 0.023). Firm *size* and *ROA* also appear to vary

Table 5.4 Univariate analysis by subsample

Panel 4A. Subsamples based on Family ownership						
Variables	Low family ownership			High family ownership			Two-tailed t-test for difference in means
	Obs	Mean	Std. dev.	Obs	Mean	Std. dev.	p-value
Innovation scale	166	1.05	2.83	234	2.17	6.51	0.019
Innovation basicness	166	8.46	29.05	234	21.88	86.33	0.027
Innovation value	166	1.09	4.14	234	1.50	5.94	0.220
Technological scope	166	1.77	5.52	234	3.73	11.76	0.023
Employee	166	365.44	382.79	215	286.51	334.72	0.016
ROA	166	9.27	8.99	223	6.82	9.29	0.005

Panel 4B. Subsamples based on Family CEO						
Variables	No Family CEO			Family CEO			Two-tailed t-test for difference in means
	Obs	Mean	Std. dev.	Obs	Mean	Std. dev.	p-value
Innovation scale	164	1.57	4.88	218	1.94	5.83	0.254
Innovation basicness	164	10.77	30.12	218	21.81	6.05	0.065
Innovation value	164	0.80	3.56	218	1.84	6.39	0.031
Technological scope	164	2.70	8.34	218	3.31	10.96	0.276
Employee	161	339.73	365.75	214	314.46	354.72	0.251
ROA	162	7.98	11.26	217	8.10	7.46	0.449

(continued)

Table 5.4 (continued)

Panel 4C. Subsamples based on CEO duality						
Variables	No CEO duality			CEO Dduality		Two-tailed t-test for difference in means
	Obs	Mean	Std. dev.	Obs	Mean	Std. dev.	p-value
Innovation scale	278	1.75	5.78	104	1.87	4.41	0.426
Innovation basicness	278	10.54	31.54	104	34.51	123.44	0.001
Innovation value	278	1.11	4.66	104	2.15	6.92	0.046
Technological scope	278	3.03	10.67	104	3.09	7.56	0.480
Employee	272	319.09	336.06	103	341.73	415.61	0.293
ROA	276	8.21	10.04	103	7.61	6.76	0.288

Panel 4D. Subsamples based on Family Supervision						
Variables	Low family supervision			High family supervision		Two-tailed t-test for difference in means
	Obs	Mean	Std. dev.	Obs	Mean	Std. dev.	p-value
Innovation scale	186	1.47	3.71	214	1.91	6.41	0.206
Innovation basicness	186	20.71	90.86	214	12.48	41.07	0.117
Innovation value	186	1.23	5.22	214	1.42	5.32	0.360
Technological scope	186	2.38	6.05	214	3.38	12.01	0.152
Employee	183	365.49	349.85	198	279.69	361.62	0.009
ROA	183	7.98	9.33	206	7.76	9.16	0.409

Notes To define low and high family supervision (*family ownership*), this study uses firms' mean value for family supervision (*family ownership*) considerably across firms with low and high ownership, such that the last subsample is composed of smaller companies (286.51 versus 365. 44, $p = 0.016$) with lower performance (6.82 versus 9.27, $p = 0.005$).

A univariate comparison also can investigate differences in means among firms with different leadership structures. Panels 4B and 4C present the subsample differences according to *family CEO* and *CEO duality*. With regard to the role of *family CEO*, the results exhibit no significant differences in terms of *innovation scale* or *technological scope* and indicate no differences in *size* and *ROA*. Interestingly, firms run by a *family CEO* exhibit significantly higher *innovation basicness* (21.81 versus 10.77, $p = 0.065$) and *value* (1.84 versus 0.80, $p = 0.031$). Panel 4C illustrates a similar pattern for firms that have a unity-of-command leadership structure: When the firm's CEO also holds the position of board chair, both *innovation basicness* and *innovation value* are, on average, significantly higher (34.51 vs. 10.54, $p = 0.001$; 2.15 versus 1.11, $p = 0.046$, respectively).

Finally, Table 5.4, Panel D illustrates differences across firms with low (below the mean value) and high (above the mean value) *family supervision*. Although the difference in means is not significant, firms with high *family supervision* show a higher *innovation scale* (1.91 versus 1.47, $p = 0.206$). With regard to qualitative dimensions of innovation performance, the univariate statistics do not show significant differences between low- and high-supervision firms. However, the difference in means suggests that firms with lower *family supervision* of their boards of directors have higher *innovation basicness* but lower *innovation value* and *technological scope*. Panel 4D also reports that family firms with lower supervision of boards of directors are significantly larger in terms of number of employees (365.49 vs. 279.69, $p = 0.001$).

Collectively, this univariate comparison of the family firm types included in this study sample yields preliminary indications that they differ substantially with regard to technological performance. However, these results should be interpreted with caution, because they do not control for other differences in firm characteristics. They thus suggest the need to account not only for the multidimensionality that characterizes family influence, as clearly developed in prior family business studies, but also to unpack innovation performance and shed light on the puzzling effect of family firms on both the scale and quality of innovation.

5.5   Findings

Table 5.5 reports the results of the negative binomial regression analysis that explores the relationship between family governance variables and the quantitative dimension of firms' technological innovation performance. To select the most appropriate specification between fixed and random-effects models, a Hausman test is used on the full model, yielding statistically significant results of $\chi^2 = 92.06$ ($p < 0.000$; Model 4, Table 5.5). Therefore, the Hausman test rejects the null hypothesis, indicating that the random-effect estimator is inconsistent and a fixed-effect estimator is appropriate. In such a case, firm fixed effects help control for any time-invariant, unobserved, firm-level variables that may influence the results. In addition, using the fixed-effects specification reduces the possibility of another source of endogeneity, i.e., omitted variables.

Model 1 tests the baseline model, including the control variables only. The results highlight a significant and positive effect of firm *size* ($\beta = 0.00$, $p < 0.05$) indicating that larger firms have a higher *innovation scale*, and a negative and significant effect of the firm's innovative *experience* ($\beta = -0.09$, $p < 0.01$), which suggests that firms that have started to innovate more recently are also more prolific in terms of patent production. None of the other controls is significant. Model 2 includes *family ownership*, which has a positive but nonsignificant effect on *innovation scale*. Model 3 includes the effects of *family CEO* and *CEO duality*, highlighting no significant effect of the leadership structure: *family CEO* has a positive but nonsignificant effect, and *CEO duality* shows a nonsignificant negative effect. In Model 4, *family supervision* is added, and it exhibits a positive and significant coefficient ($\beta = 0.68$, $p < 0.05$). The greater involvement of family members in boards of directors seems to influence *innovation scale* significantly and positively.

This first set of models points to two interesting findings. First, only family supervision plays a decisive role in *innovation scale*. Second, the

Table 5.5 Negative binomial regression models for innovation scale

Innovation Scale	Model 1 Negative binomial	Model 2 Negative binomial	Model 3 Negative binomial	Model 4 Negative binomial
Collaboration	0.16	0.08	0.08	0.06
	(0.20)	(0.20)	(0.21)	(0.21)
Experience	−0.09***	−0.12***	−0.12***	−0.14***
	(0.03)	(0.03)	(0.03)	(0.04)
Technological diversity	−0.27	−0.24	−0.26	−0.36
	(0.26)	(0.28)	(0.28)	(0.29)
Patent stock	0.28	0.21	0.20	0.08
	(0.22)	(0.23)	(0.23)	(0.24)
ROA	0.00	−0.01	−0.00	−0.01
	(0.02)	(0.02)	(0.02)	(0.02)
Size	0.00**	0.00***	0.00***	0.00***
	(0.00)	(0.00)	(0.00)	(0.00)
Family ownership		0.01	0.01	0.01
		(0.01)	(0.01)	(0.01)
Family CEO			0.09	0.40
			(0.22)	(0.26)
CEO duality			−0.09	−0.56
			(0.38)	(0.43)
Family supervision				1.68**
				(0.80)
Constant	3.32***	3.42***	3.48***	3.86***
	(1.13)	(1.19)	(1.29)	(1.34)
Fixed effect	FE	FE	FE	FE
Number of observations	152	138	138	138
Wald χ^2	24.97	30.14	30.72	35.55
Prob. $> \chi^2$	0.00	0.00	0.00	0.00
Log likelihood	−182.48	−168.91	−168.83	−166.71

Notes Standard errors are in parentheses
*$p < 0.1$, **$p < 0.05$, ***$p < 0.01$

effect of *family supervision,* i.e., the presence of family members in the board of directors, on *innovation scale* is positive. Thus, contrary to the expectations of conventional theoretical models, in the high-technology context of the pharmaceutical industry, greater involvement of family members in boards of directors seems to generate a significantly greater number of patents. As suggested above, a possible explanation for this finding is that family members involved in boards of directors realize that the specific features of the industry make innovation a vital tool for staying abreast of industry technology and strategically adapting to rapidly changing external environments—both of which preserve family firms in the long run. In other words, consistent with predictions of the BAM, socio-emotional attachment drives family board members to exercise their monitoring and guidance roles to generate the amount of innovations necessary to survive in a high-technology industry, even though innovation is a risky activity. The validity of this finding seems to be supported by qualitative data gathered during interviews with two family affiliates who are both owners and board members of a major Italian pharmaceutical company, who recognize that maintaining a rich patent portfolio is a precondition of survival in the industry. They point out that their family-influenced board of directors advises managers to make critical trade-offs between long-term investments (such as R&D projects) and short-term financial results. They recognize the broad and pivotal role of the board of directors, not only in monitoring and controlling company managers, but also in advising managers on crucial issues such as innovation activity. For instance, they state that—in spite of their lack of technical competence in the pharmaceutical field—they are very receptive to the technological changes that occur in the industry and encourage the firm R&D manager to continuously monitor the wider technological landscape to ensure the firm is aware of any significant advancement in the industry knowledge-base.

In line with the aforementioned arguments on the multidimensional nature of innovation performance, this study also investigates the quality of technological innovation by unpacking it into several dimensions, namely, *innovation basicness*, *innovation value*, and *technological scope*.

Table 5.6 depicts the relationship between *family governance* variables and *innovation basicness*. For this model, a Hausman test yields no statistically significant results of $\chi^2 = 1.35$ ($p < 0.993$; Model 8A versus Model 8B, Table 5.6), thus advising toward the use of a random-effects specification. The following discussion thus focuses on the results of random-effects models, although a fixed- effects specification is also reported (see Model 8B), showing that the main results are robust across

Table 5.6 Negative binomial regression models for innovation basicness

Innovation basicness	Model 5 Negative binomial	Model 6 Negative binomial	Model 7 Negative binomial	Model 8A Negative binomial	Model 8B Negative binomial
Collaboration	0.14	0.14	0.29	0.09	−0.08
	(0.20)	(0.20)	(0.19)	(0.17)	(0.20)
Experience	0.03**	0.03**	0.02*	0.02*	−0.04*
	(0.01)	(0.01)	(0.01)	(0.01)	(0.02)
Technological diversity	−1.08***	−1.04***	−1.09***	−0.95***	−0.29
	(0.28)	(0.29)	(0.29)	(0.28)	(0.32)
Patents	1.36***	1.33***	1.36***	1.57***	1.35***
	(0.10)	(0.10)	(0.11)	(0.11)	(0.11)
ROA	−0.01	−0.01	−0.02	−0.02	−0.05**
	(0.02)	(0.02)	(0.02)	(0.02)	(0.02)
Size	0.00*	0.00*	0.00*	0.00**	0.00***
	(0.00)	(0.00)	(0.00)	(0.00)	(0.00)
Family ownership		−0.00	0.00	−0.00	0.01
		(0.00)	(0.00)	(0.01)	(0.01)
Family CEO			0.43**	0.32*	0.40*
			(0.21)	(0.19)	(0.22)
CEO duality			0.05	0.50*	0.08
			(0.29)	(0.30)	(0.34)
Family supervision				−2.29***	−2.16***
				(0.46)	(0.63)
Constant	−2.66***	−2.48*	−2.81***	−1.83***	−1.38
	(0.35)	(0.61)	(0.65)	(0.68)	(0.85)
Fixed effect–Random effect	RE	RE	RE	RE	FE
Number of observations	331	294	294	294	138
Wald χ²	439.21	407.83	407.86	483.59	289.63
Prob. > χ²	0.00	0.00	0.00	0.00	0.00
Log likelihood	−473.89	−452.44	−449.10	−435.43	−295.96

Notes Standard errors are in parentheses
*$p < 0.1$, **$p < 0.05$, ***$p < 0.01$

different estimations and are consistent at both cross-sectional and within-firm levels.

Model 5 tests the baseline model, including control variables. As expected, a significant and positive effect emerges for the (log of the) number of patents applied for in year t ($\beta = 1.36$, $p < 0.01$), since a higher number of patents encompasses a higher likelihood of referencing scientific citations. Also firm *size* ($\beta = 0.00$, $p < 0.10$) and *experience* ($\beta = 0.03$, $p < 0.05$) are positively and significantly associated with *innovation basicness*, suggesting that larger firms and firms that have been involved in innovative activities for a longer period of time tend to generate more basic knowledge. On the contrary, the results reveal a negative relation between *technological diversity* and *innovation basicness* ($\beta = -1.08$, $p < 0.01$); thus, firms that work on a less technologically focused pipeline tend to rely less on scientific knowledge. Model 6 confirms a nonsignificant effect of *family ownership*. Model 7 reveals a positive and significant effect of *family CEO* on *innovation basicness* ($\beta = 0.43$, $p < 0.05$); *CEO duality* shows a nonsignificant effect. In Model 8A, *family supervision* is added to capture the influence of family board members. The result indicates that greater involvement of family members in the board of directors negatively and significantly influences the firm's *innovation basicness* ($\beta = -2.29$, $p < 0.01$).

This second set of models confirms that *family supervision* also has a decisive role for *innovation basicness*. However, contrary to findings about the quantitative dimension of innovation, these results show that in the pharmaceutical industry, greater involvement of family firms in boards of directors generate innovations with significantly less basicness—an interesting and powerful result. The basicness of technological innovation could be associated with very hazardous inventive activities that use fundamental knowledge to push the technological frontier. A substantial leverage of fundamental knowledge usually generates innovations that are distant from bodies of practice, address general technological problems, and have limited commercialization potential, at least in the short term (Kuhn 1962; Rosenberg 1982). Therefore, these innovations may be perceived as riskier, both because their commercial outcomes are less certain and because their technological implications are more difficult to recognize and to be applied for practical objectives. Consistent with the theoretical expectations of BAMs, these results suggest that even in contexts in which innovation is an inevitable component of a firm's activities, once the long-term aim of firm survival is ensured through a significant scale of innovation, the family's presence on the board of directors serves a monitoring function, to pursue inherent risk aversion. As a consequence, the greater the family's involvement in the board of directors, the lower the firm's engagement in innovative activities that leverage basic and fundamental knowledge to spur innovation related to bodies of science. Although these innovations may embed more radical improvements of the existing technology base, they are also perceived as very risky; thus family-dependent boards of directors discourage their pursuit to protect the family SEW. Interestingly, this result seems to be validated by qualitative data gathered through interviews with the manager of a non-family firm operating in the Italian pharmaceutical industry. Speaking about the family board members of an important Italian family firm, this informant expressed disapproval over their approach to the Italian scientific world and stated that if they were interested in exploring truly ambitious innovative projects, they would have invested more in knowledge sourcing from the local universities and research centers. Conversely, they prefer to conduct more traditional, in-house innovative activities—a course of action that, in the past, has prevented them to exploit technological opportunities arising from the Italian scientific system of innovation, which paradoxically have been captured by a foreign firm.

This set of models also documents a positive relationship between *family CEO* and *innovation basicness*. This result seems to show that not all dimensions of family involvement in a firm's business have the same effect on the firm's technological performance. In this case, *family supervision* and *family management* push in opposite directions: The presence of family directors drives the firm to lower its engagement in basic innovative activities, but the involvement of a family member as a CEO pushes it to use scientific knowledge to generate fundamental innovations. These results point to an active role of family management, whereby family managers operating in contexts that feature a high intensity of technology comprehensively understand the industry profile and competitive dynamics and—given their personal commitment to the firm's growth—adopt a long-term approach to the definition of the firm's goals (Maury 2006; Matzler et al. 2015). Family managers then recognize the critical importance of the bodies of science in the pharmaceutical industry and push for the systematic use of scientific knowledge to feed their firms' innovation funnels.

Such evidence can be extended to considerations of another relevant qualitative dimension of firm innovation performance: *innovation value* (Trajtenberg 1990). When estimating this dependent variable, the random-effects specification model is not significant and the Hausman test fails to meet the asymptotic assumptions; thus, Table 5.7 reports only the fixed-effects specification results to control for any time-invariant, unobserved firm-level characteristics.

Model 9 introduces control variables only, highlighting the positive and significant relationship between the number of patents successfully applied for by the firm in year t and the forward citations these patents receive ($\beta = 1.56$, $p < 0.01$) and a negative relation with *collaboration* ($\beta = -0.70$, $p < 0.10$) and *experience* ($\beta = -0.16$, $p < 0.05$), which seems to suggest that, in the Italian pharmaceutical industry, firms that collaborate on innovative activities and firms that are technologically more mature tend to generate less valuable innovations. Model 10 presents the effect of *family ownership* alone, showing a positive but nonsignificant effect ($\beta = 0.01$, $p = 0.767$). Model 11 documents the positive and significant effect of *family CEO* on the firm's *innovation value* ($\beta = 0.47$, $p < 0.05$) and the negative and significant effect of *CEO duality* ($\beta = -1.76$, $p < 0.01$). These findings are consistent with the results obtained in the specifications that estimate *innovation basicness*, pointing to the positive role that *family management* has for the quality of firm innovation. In addition to being sensitive to the generation of innovations that leverage scientific knowledge and show promise in terms of pushing the technological frontier, family CEOs are also likely to favor the development of innovations that have substantial impact and value, thus spurring more immediate economic outcomes. This result describes family CEOs as perceptive, key figures who promote wise, conducive innovation behaviors and visualize both long- and short-term innovation objectives, in an industry in which both financial outcomes and scientific discovery are crucial. As far as the leadership

Table 5.7 Negative binomial regression models for innovation value

Innovation value	Model 9 Negative binomial	Model 10 Negative binomial	Model 11 Negative binomial	Model 12 Negative binomial
Patent age	0.17	0.20	0.29***	0.41***
	(0.13)	(0.14)	(0.09)	(0.11)
Collaboration	−0.70*	−0.69*	−1.07***	−1.11***
	(0.41)	(0.42)	(0.28)	(0.30)
Experience	−0.16**	−14**	−0.11	−0.09
	(0.07)	(0.07)	(0.07)	(0.07)
Technological diversity	0.57	0.39	0.94	0.90
	(0.96)	(1.02)	(1.00)	(1.01)
Patents	1.56***	1.49***	2.38***	2.39***
	(0.38)	(0.40)	(0.40)	(0.41)
ROA	−0.13	−0.12	−0.28***	−0.29***
	(0.10)	(0.09)	(0.04)	(0.04)
Size	0.00	0.00	0.00	0.00
	(0.00)	(0.00)	(0.00)	(0.00)
Family ownership		0.01	0.01	0.02
		(0.02)	(0.02)	(0.02)
Family CEO			0.47**	0.87***
			(0.22)	(0.33)
CEO duality			−1.76**	−2.52***
			(0.74)	(0.89)
Family supervision				1.86*
				(1.09)
Constant	−1.33	−2.68	−3.67	−6.20*
	(2.64)	(3.69)	(3.37)	(3.64)
Fixed effect–Random effect	FE	FE	FE	FE
Number of observations	90	85	85	85
Wald χ^2	58.97	59.13	101.00	101.59
Prob. > χ^2	0.00	0.00	0.00	0.00
Log likelihood	−99.20	−94.52	−88.78	−87.30

Notes Standard errors are in parentheses
*$p < 0.1$, **$p < 0.05$, ***$p < 0.01$

structure is concerned, the negative and significant effect of *CEO duality* on innovation value is consistent with agency theory predictions that suggest separating the CEO and Chair of the board positions to facilitate the monitoring role of board of director (Fama 1980; Fama and Jensen 1983; Jensen 1993). Moreover, the separate leadership structure can also favor the collaboration between board and CEO, providing the CEO with a potential source of advice in managing the firm's strategy (Krause 2017). Collectively, these findings confirm the positive role of *family CEO*, but also suggest that unity of command violates the principle of independent governance (Krause and Semadeni 2013) thus exerting a negative impact on *innovation value*.

Finally, Model 12 surprisingly reveals that *family supervision* has a positive impact on *innovation value* ($\beta = 1.86$, $p < 0.10$). This finding seems to suggest that the family board members enhance the firm's involvement in innovative projects aimed at the creation of immediate commercial value. Hence, if considered in combination with the negative influence *family supervision* exerts on *innovation basicness*, the results point to a very articulated role of family board members: these agents encourage innovative projects leading to more certain economic outcomes, but dissuade truly hazardous inventive activities whose fulfillment is very unlikely. Overall, this finding is consistent with the idea that family board members are strongly committed to advise the firm's managers to take decisions that pursue the preservation of the SEW.

Finally, the analysis focuses on the *technological scope* of firms' inventive activities, which captures the extent to which a firm's innovations are relevant to various technological fields. Since random-effects estimations do not converge, fixed-effects models are used as the only possible estimation option. In Table 5.8, Model 13 reports the results of the baseline model and shows that as expected a higher number of *patents* is significantly associated with a higher number of technological classes ($\beta = 1.40$, $p < 0.01$). Interestingly, the negative and significant coefficient of *collaboration* ($\beta = -0.26$, $p < 0.05$) suggests that, in the Italian pharmaceutical industry, collaborating for innovation tends to be associated with a narrower technological scope of a firm patent production. Because of problems of model convergence, it was not possible to estimate a model adding only family ownership to the baseline model.

Table 5.8 Negative binomial regression models for technological scope

Technological scope	Model 13 Negative binomial	Model 14 Negative binomial	Model 15 Negative binomial
Collaboration	−0.26**	−0.27**	−0.27**
	(0.11)	(0.11)	(0.11)
Experience	0.02	0.03	0.02
	(0.02)	(0.02)	(0.03)
Technological diversity	−0.25	−0.18	−0.17
	(0.20)	(0.22)	(0.22)
Patents	1.40***	1.37***	1.40***
	(0.10)	(0.10)	(0.11)
ROA	−0.01	−0.01	−0.01
	(0.01)	(0.01)	(0.01)
Size	0.00	−0.00	−0.00
	(0.00)	(0.00)	(0.00)
Family ownership		0.00	0.00
		(0.01)	(0.00)
Family CEO		−0.14	−0.21
		(0.11)	(0.14)
CEO duality		−0.07	0.04
		(0.24)	(0.27)
Family supervision			−0.43
			(0.51)
Constant	14.34	15.75	13.69
	(874.76)	(703.88)	(348.33)
Fixed effect–Random effect	FE	FE	FE
Number of observations	152	152	138
Wald χ^2	236.06	227.51	228.61
Prob. $> \chi^2$	0.00	0.00	0.00
Log likelihood	−163.22	−151.46	−151.10

Notes Standard errors are in parentheses
$^*p < 0.1$, $^{**}p < 0.05$, $^{***}p < 0.01$

Hence, Model 14 includes both the ownership variable and the variables depicting the firm's leadership structure. None of these variables turn out to be significant. Model 15 also includes family supervision, but also in this case, no significant effects are reported. Thus, it seems that family influence does not exert any role on this dimension of innovation quality. One possible explanation is that family members lack sufficient technical competence to understand the mechanisms linking technological scope to the firm's SEW, thus being unable to exert significant influence on this aspect.

Collectively, the empirical findings reveal that the influence of the family on a firm's innovation output is a multidimensional phenomenon, since different dimensions of the family firms may heterogeneously affect distinct aspects of the firm's innovation output.

5.6   Sensitivity Analysis and Robustness Checks

As explained below, a number of tests have been conducted to ensure the robustness of results with regard to issues of multicollinearity and endogeneity.

Due to high correlations between some of the variables employed in this analysis, several sensitivity analyses have been conducted. First, with regard to the specification estimating the *scale of innovation*, the full model was run excluding—respectively—*family CEO, family supervision, patent stock,* firm *size,* and *experience.* Overall, the results confirm the findings described above, except for the specifications that exclude *family CEO* and *experience,* when the positive effect of *family supervision* becomes marginally nonsignificant. Second, with regard to specifications estimating the qualitative dimensions of innovation performance (*value, basicness,* and *technological scope*), the full models were run excluding—respectively—*family CEO, family supervision,* firm *size,* and *experience.* The main findings of the analysis are confirmed, with a few exceptions. First, the positive effect of *family CEO* on the *value* of innovation becomes marginally nonsignificant when removing firm *size.* Second, the positive effect of *family supervision* disappears when *family CEO* is removed from the full model. Hence, the latter finding should be taken with caution and further tested by future research.

Moreover, since the *technological diversity* variable has been set equal to 1 for firm-year combinations in which our sample firms did not have any patent to avoid dropping observations, we re-run our models excluding this variable. All our main results are confirmed with the exception of the positive and significant effect of *family CEO* on the *basicness* of innovation, which becomes nonsignificant in the full model, although it stays positive and significant when testing only the effect of the leadership structure (i.e. when *family ownership* and *family supervision* are removed from the model).

To control for potential endogeneity issues associated with family-managed firms, in line with previous work on family business (e.g., Villalonga and Amit 2006; Miller et al. 2013; Block et al. 2013), this study sought to estimate a set of two-step treatment effects models, in which the dependent variable of the first stage is *family CEO* and the dependent variables of the second stage are the various count variables of interest (*innovation scale, innovation basicness, innovation value,* and *technological scope*). In the first stage, in addition to the control variables used in the different baseline models, specifically ROA and firm size, two variables are included: (1) *family ownership*, which accounts for the presence of family members and their influence on the decision to appoint a *family CEO* (Miler et al. 2013) and (2) the amount of *cash holding*, defined as the natural logarithm of the ratio between cash and market securities divided by total revenues, which functions as an instrument to identify the *family CEO* (Miller et al. 2007). The second stage relies on the predicted value of *family CEO* to estimate the various innovation performance variables of interests. Finally, in the two-step treatment effects model, robust standard errors are clustered at the firm level. A test of the model that estimates *innovation scale* confirms the effects of governance variables (and specifically the negative and significant effect of *family supervision*) while the model for *innovation value* is not significant.[3] Unfortunately, problems of model convergence prevented to use this specification to re-estimate *innovation basicness* or *technological scope*.

Notes

1. Three specialized industry consultants, two family owners who also participate on their firms' boards of directors, one family manager, and two non-family managers operating in the Italian pharmaceutical industry were interviewed during the data collection process.

2. Data is available at the following https://sites.google.com/site/mcgoossen/patent-grant.
3. In particular, the Wald tests of independent equations indicate a not significant correlation between the error terms of the first and second stage equations, suggesting the absence of selectivity bias; hence, the use of treatment effect models is not justified.

References

Achilladelis, B., & Antonakis, N. (2001). The dynamics of technological innovation: The case of the pharmaceutical industry. *Research Policy, 30*(4), 535–588.

Almeida, P., & Phene, A. (2004). Subsidiaries and knowledge creation: The influence of the MNC and host country on innovation. *Strategic Management Journal, 25*(8–9), 847–864.

Anderson, R. C., Duru, A., & Reeb, D. M. (2012). Investment policy in family controlled firms. *Journal of Banking & Finance, 36*(6), 1744–1758.

Bammens, Y., Voordeckers, W., & Van Gils, A. (2011). Boards of directors in family businesses: A literature review and research agenda. *International Journal of Management Reviews, 13*(2), 134–152.

Banno, M. (2016). Propensity to patent by family firms. *Journal of Family Business Strategy, 7*(4), 238–248.

Basu, N., Dimitrova, L., & Paeglis, I. (2009). Family control and dilution in mergers. *Journal of Banking & Finance, 33*(5), 829–841.

Bierly, P., & Chakrabarti, A. (1996). Generic knowledge strategies in the U.S. pharmaceutical industry. *Strategic Management Journal, 17,* 123–135.

Block, J. H. (2012). R&D investments in family and founder firms: An agency perspective. *Journal of Business Venturing, 27*(2), 248–265.

Block, J., Miller, D., Jaskiewicz, P., & Spiegel, F. (2013). Economic and technological importance of innovations in large family and founder firms: An analysis of patent data. *Family Business Review, 26*(2), 180–199.

Bloom, N., & Van Reenen, J. (2002). Patents, real options and firm performance. *The Economic Journal, 112*(478), 97–116.

Broekaert, W., Andries, P., & Debackere, K. (2016). Innovation processes in family firms: The relevance of organizational flexibility. *Small Business Economics, 47*(3), 771–785.

Carnes, C. M., & Ireland, R. D. (2013). Familiness and innovation: Resource bundling as the missing link. *Entrepreneurship Theory and Practice, 37*(6), 1399–1419.

Cassiman, B., Veugelers, R., & Zuniga, P. (2008). In search of performance effects of (in) direct industry science links. *Industrial and Corporate Change, 17*(4), 611–646.

Chen, H. L., & Hsu, W. T. (2009). Family ownership, board independence, and R&D investment. *Family Business Review, 22*(4), 347–362.

Chesbrough, H. (2006). Open innovation: A new paradigm for understanding industrial innovation. *Open Innovation: Researching a New Paradigm, 400,* 0–19.

Chrisman, J. J., & Patel, P. C. (2012). Variations in R&D investments of family and nonfamily firms: Behavioral agency and myopic loss aversion perspectives. *Academy of Management Journal, 55*(4), 976–997.

Classen, N., Carree, M., Van Gils, A., & Peters, B. (2014). Innovation in family and non-family SMEs: An exploratory analysis. *Small Business Economics, 42*(3), 595–609.

Cockburn, I. M., & Henderson, R. M. (2000). Publicly funded science and the productivity of the pharmaceutical industry. *Innovation Policy and the Economy, 1,* 1–34.

Corbetta, G., & Salvato, C. A. (2004). The board of directors in family firms: One size fits all? *Family Business Review, 17*(2), 119–134.

DeCarolis, D. M. (2003). Competencies and imitability in the pharmaceutical industry: An analysis of their relationship with firm performance. *Journal of Management, 29,* 27–50.

Duran, P., Kammerlander, N., Van Essen, M., & Zellweger, T. (2016). Doing more with less: Innovation input and output in family firms. *Academy of Management Journal, 59*(4), 1224–1264.

Fama, E. F. (1980). Agency problems and the theory of the firm. *The Journal of Political Economy, 88*(2), 288–307.

Fama, E. F., & Jensen, M. C. (1983). Agency problems and residual claims. *Journal of Law and Economics, 26,* 327–349.

Family Firms Institute. (2017). *Family enterprise statistics from around the world.* Available at: http://www.ffi.org/?page=GlobalDataPoints.

Gambardella, A. (1992). Competitive advantages from in-house scientific research: The US pharmaceutical industry in the 1980s. *Research Policy, 21*(5), 391–407.

Gambardella, A. (2005). Patents and the division of innovative labor. *Industrial and Corporate Change, 14*(6), 1223–1233.

Gambardella, A., Harhoff, D., & Verspagen, B. (2008). The value of European patents. *European Management Review, 5*(2), 69–84.

Gittelman, M., & Kogut, B. (2003). Does good science lead to valuable knowledge? Biotechnology firms and the evolutionary logic of citation patterns. *Management Science, 49*(4), 366–382.

Gomez-Mejia, L. R., Campbell, J. T., Martin, G., Hoskisson, R. E., Makri, M., & Sirmon, D. G. (2014). Socioemotional wealth as a mixed gamble: Revisiting family firm R&D investments with the behavioral agency model. *Entrepreneurship Theory and Practice, 38*(6), 1351–1374.

Gomez-Mejia, L. R., Haynes, K. T., Nunez-Nickel, M., Jacobson, K., & Moyano-Fuentes, J. (2007). Socioemotional wealth and business risks in family controlled firms: Evidence from Spanish olive oil mills. *Administrative Science Quarterly, 52,* 106–137.

Hagedoorn, J. (2003). Sharing intellectual property right: An exploratory study of joint patenting amongst companies. *Industrial and Corporate Change, 12,* 1035–1050.

Hall, B. R. H., & Ziedonis, A. A. (2001). The patent paradox revisited: An empirical study of patenting in the US. *The Rand Journal of Economics, 32*(1), 101–128.

Hall, B. H., Jaffe, A. B., & Trajtenberg, M. (2001). *The NBER patent citation data file: Lessons, insights and methodological tools (No. w8498).* National Bureau of Economic Research.

Hall, B. H., Jaffe, A., & Trajtenberg, M. (2005). Market value and patent citations. *Rand Journal of Economics,* 16–38.

Hess, A. M., & Rothaermel, F. T. (2011). When are assets complementary? Star scientists, strategic alliances, and innovation in the pharmaceutical industry. *Strategic Management Journal, 32*(8), 895–909.

Hsu, P., Huang, S., Massa, M., & Zhang, H. (2014). *The new lyrics of the old folks: The role of family ownership in corporate innovation* (Working paper). INSEAD.

Jaffe, A. B., Trajtenberg, M., & Henderson, R. (1993). Geographic localization of knowledge spillovers as evidenced by patent citations. *The Quarterly Journal of Economics, 108*(3), 577–598.

Jensen, M. C. (1993). The modern industrial revolution, exit, and the failure of internal control systems. *The Journal of Finance, 48*(3), 831–880.

Jensen, M. C., & Meckling, W. H. (1976). Theory of the firm: Managerial behavior, agency costs and ownership structure. *Journal of Financial Economics, 3*(4), 305–360.

Kraiczy, N. D., Hack, A., & Kellermanns, F. W. (2015). What makes a family firm innovative? CEO risk-taking propensity and the organizational context of family firms. *Journal of Product Innovation Management, 32*(3), 334–348.

Kraiczy, N. D., Hack, A., & Kellermanns, F. W. (2014). New product portfolio performance in family firms. *Journal of Business Research, 67*(6), 1065–1073.

Krause, R. (2017). Being the CEO's boss: An examination of board chair orientations. *Strategic Management Journal, 38,* 697–713.

Krause, R., & Semadeni, M. (2013). Apprentice, departure, and demotion: An examination of the three types of CEO-board chair separation. *Academy of Management Journal, 56*(3), 805–826.

Kuhn, T. S. (1962). *The structure of scientific revolutions.* Chicago University Press.

La Porta, R., Lopez-de-Silanes, F., & Shleifer, A. (1999). Corporate ownership around the world. *Journal of Finance, 54*(2), 471–517.

Le Breton-Miller, I., Miller, D., & and Lester, R. H. (2011). Stewardship or agency? A social embeddedness reconciliation. *Organization Science, 22*(3), 704–721.

Lee, J., & Berente, N. (2012). Digital innovation and the division of innovative labor: Digital controls in the automotive industry. *Organization Science, 23*(5), 1428–1447.

Lerner, J. (1994). The importance of patent scope: An empirical analysis. *The Rand Journal of Economics,* 319–333.

Levin, R. C., Klevorick, A. K., Nelson, R. R., Winter, S. G., Gilbert, R., & Griliches, Z. (1987). Appropriating the returns from industrial research and development. *Brookings Papers on Economic Activity, 3,* 783–831.

Lodh, S., Nandy, M., & Chen, J. (2014). Innovation and family ownership: Empirical evidence from India. *Corporate Governance—An International Review, 22*(1), 4–23.

Matzler, K., Veider, V., Hautz, J., & Stadler, C. (2015). The impact of family ownership, management, and governance on innovation. *Journal of Product Innovation Management, 32*(3), 319–333.

Maury, B. (2006). Family ownership and firm performance: Empirical evidence from Western European corporations. *Journal of Corporate Finance, 12*(2), 321–341.

McConaughy, D. L., Matthews, C. H., & Fialko, A. S. (2001). Founding family controlled firms: Performance, risk, and value. *Journal of Small Business Management, 39*(1), 31–49.

Miller, D., Minichilli, A., & Corbetta, G. (2013). Is family leadership always beneficial? *Strategic Management Journal, 34*(5), 553–571. doi:10.1002/smj.2024.

Miller, D., Le Breton-Miller, I., Lester, R. H., & Cannella, A. A. (2007). Are family firms really superior performers?. *Journal of Corporate Finance, 13*(5), 829–858.

Mishra, C. S., & McConaughy, D. C. (1999). Founding family control and capital structure: The risk of loss of control and the aversion to debt. *Entrepreneurship Theory and Practice, 23*, 53–65.

Munari, F., Oriani, R., & Sobrero, M. (2010). The effects of owner identity and external governance systems on R&D investments: A study of western european firms, *Research Policy, 39*(8), 1093–1104.

Nerkar, A., & Paruchuri, S. (2005). Evolution of R&D capabilities: The role of knowledge networks within a firm. *Management Science, 51*(5), 771–785.

Nieto, M. J., Santamaria, L., & Fernandez, Z. (2015). Understanding the innovation behavior of family firms. *Journal of Small Business Management, 53*(2), 382–399.

OECD. (2011). Technology-science linkages. In *OECD science, technology and industry scoreboard 2011*. OECD Publishing. http://dx.doi.org/10.1787/sti_scoreboard-2011-25-en.

Patel, P. C., & Chrisman, J. J. (2014). Risk abatement as a strategy for R&D investments in family firms. *Strategic Management Journal, 35*, 617–627.

Penner-Hahn, J., & Shaver, J. M. (2005). Does international research and development increase patent output? An analysis of Japanese pharmaceutical firms. *Strategic Management Journal, 26*(2), 121–140.

Perri, A., & Andersson, U. (2014). Knowledge outflows from foreign subsidiaries and the tension between knowledge creation and knowledge protection: Evidence from the semiconductor industry. *International Business Review, 23*(1), 63–75.

Perri, A., Scalera, V. G., & Mudambi, R. (2017). What are the most promising conduits for foreign knowledge inflows? Innovation networks in the Chinese pharmaceutical industry. *Industrial and Corporate Change, 26*(2), 333–355.

Phene, A., & Almeida, P. (2008). Innovation in multinational subsidiaries: The role of knowledge assimilation and subsidiary capabilities. *Journal of International Business Studies, 39*, 901–919.

Rosenberg, N. (1982). *Inside the black box: Technology and economics*. Cambridge University Press.

Scalera, G. V., Mukherjee, D., Perri, A., & Mudambi, R. (2014). A longitudinal study of MNE innovation: The case of Goodyear. *Multinational Business Review, 22*(3), 270–293.

Schmid, T., Achleitner, A. K., Ampenberger, M., & Kaserer, C. (2014). Family firms and R&D behavior–New evidence from a large-scale survey. *Research Policy, 43*(1), 233–244.

Schmid, T., Ampenberger, M., Kaserer, C., & Achleitner, A. K. (2015). Family firm heterogeneity and corporate policy: Evidence from diversification decisions. *Corporate Governance: An International Review, 23*(3), 285–302.

Schulze, W. S., Lubatkin, M. H., Dino, R. N., & Buchholtz, A. K. (2001). Agency relationships in family firms: Theory and evidence. *Organization Science, 12*(2), 99–116.

Sirmon, D. G., & Hitt, M. A. (2003). Managing resources: Linking unique resources, management, and wealth creation in family firms. *Entrepreneurship Theory and Practice, 27*(4), 339–358.

Spriggs, M., Yu, A., Deeds, D., & Sorenson, R. L. (2013). Too many cooks in the kitchen: Innovative capacity, collaborative network orientation, and performance in small family businesses. *Family Business Review, 26*(1), 32–50.

Teece, D. J., Pisano, G., & Shuen, A. (1997). Dynamic capabilities and strategic management. *Strategic Management Journal, 18*(7), 509–533.

Tijssen, R. J. (2001). Global and domestic utilization of industrial relevant science: Patent citation analysis of science–technology interactions and knowledge flows. *Research Policy, 30,* 35–54.

Trajtenberg, M. (1990). A penny for your quotes: Patent citations and the value of innovations. *The Rand Journal of Economics,* 172–187.

Trajtenberg, M., Henderson, R., & Jaffe, A. (1997). University versus corporate patents: A window on the basicness of invention. *Economics of Innovation and New Technology, 5*(1), 19–50.

Vanhaverbeke, W., Chesbrough, H., & West, J. (2014). Surfing the new wave of open innovation research. *New Frontiers in Open Innovation,* 281.

Villalonga, B., & Amit, R. (2006). How do family ownership, control and management affect firm value? *Journal of Financial Economics, 80*(2), 385–417.

Villalonga, B., & Amit, R. (2009). How are US family firms controlled? *Review of Financial Studies, 22,* 3047–3091.

Wadhwa, A., & Kotha, S. (2006). Knowledge creation through external venturing: Evidence from the telecommunications equipment manufacturing industry. *Academy of Management Journal, 49*(4), 819–835.

Wiseman, R. M., & Gomez-Mejia L. M. (1998). A behavioral agency model of managerial risk taking. *The Academy of Management Review, 23*(1), 133–153.

6

Concluding Remarks and Avenues for Future Research

Abstract This chapter summarizes the most relevant findings of this volume and discusses the main contributions for both family business and innovation research streams. It proposes an integrative framework that accounts for the multifaceted relationship linking family firms and innovation output. The main limitations of this book are also discussed to open up a research agenda and stimulate future research into this promising area of inquiry.

Keywords Family firms · Innovation output · Multitheoretic approach Institutional and regulatory framework

6.1 Concluding Remarks and Discussion

Family firms have been traditionally identified as conservative and risk-adverse organizations, with limited willingness to invest in innovative activities (Gomez-Mejia et al. 2007). This characterization, though, contrasts with evidence showing that many of the most success-ful, enduring, and innovative companies worldwide are family firms.

© The Author(s) 2017
A. Perri and E. Peruffo, *Family Business and Technological Innovation*,
DOI 10.1007/978-3-319-61596-7_6

This book has attempted to shed light on this puzzling phenomenon, by contributing to the understanding of the role of the family in firm innovation performance. Empirically, this book has explored a specific type of industry settings, i.e., those featuring high-technology intensity, with a focus on the Italian pharmaceutical industry. Given their fast-changing dynamics, these environments are likely to pose even more significant challenges to organizations that operate to pursue the status quo and, hence, deserve distinct research attention (Gomez-Mejia et al. 2014).

A critical assumption of this book is that a meaningful analysis of the role of the family in the output of firm innovation processes should account for both the heterogeneity of the family influence on a firm's business and the various dimensions of firm innovation performance.

On the one hand, while family firms are the most common and widely studied ownership structure (De Massis et al. 2015), the diversity of theoretical and empirical perspectives through which they have been investigated leaves the definition of a family firm still far from settled (Siebels and Knyphausen-Aufseß 2012). This book moves beyond this definition dilemma to leverage the idea that the heterogeneity of family firms displays in the multiple roles that family members can play within the company. Specifically, since firms' innovative behavior may vary depending on whether they are merely owned, controlled, or also managed by family members (e.g., Block 2012; Block et al. 2013; Matzler et al. 2015), this study unpacks the family construct to investigate the governance structures adopted by family firms, reflecting distinct alignments of ownership, control, and management (Carney 2005).

On the other hand, while research into innovation inputs tends to be more unanimous in documenting a negative influence of the family firm on the level of R&D investments, the stream on innovation outputs is more inconclusive and, even in this research subfield, significant discrepancy persists regarding the distinct roles of the various dimensions of family influence. This nascent evidence, coupled with established research that has documented the existence of significant variation in the intrinsic qualities of innovations (e.g., Trajtenberg 1990; Trajtenberg et al. 1997; Hall et al. 2001; Gambardella et al. 2008) and in firms' choice to commit to distinct dimensions of the innovative activity (Gambardella 2005; Lee and Berente 2012), suggests to adopt a fine-grained approach to the analysis of firms' technological performance.

The results of this study are manifold and offer different insights to existing research. First, in this study setting, *family ownership* does not seem to have any effect on firm innovation performance. This result is in line with prior work on the relation between family and innovation output (Matzler et al. 2015) and confirms that the effect of familiness, intended as the unique combination of resources and capabilities that generate advantage-based rents and high levels of value creation, is strictly related to the active involvement of the family in the management and governance of the firm (Maury 2006). Moreover, this is clearly a finding that should be contextualized to account for the widespread family ownership that characterizes the Italian industrial system: not only are family-owned firms ubiquitous in Italy, they also tend to feature similar degrees of family ownership, thereby reducing the heterogeneity of this dimension. Interestingly enough, this study shows that the diffusion of family-owned firms is considerable also in high-technology contexts where, in principle, many of the features that are traditionally credited to family firms (e.g., path-dependency, continuity, risk-aversion) reduce the fit with the environmental context.

Second, *family management* seems to have a positive effect on the qualitative dimensions of firm innovation, but does not influence its *scale*. This result is consistent with traditional agency theory (e.g., Jensen and Meckling 1976), with the resource-based view of the family (e.g., Habberson and Williams 1999), and with subsequent stewardship perspectives (e.g., Davis et al. 1997). In fact, unlike non-family managers whose behavior typically focuses on short-term performance, family managers reduce agency costs (Dyer 2006), have a long-term orientation (Matzler et al. 2015), and may also leverage a whole set of familial assets, including a significant knowledge of the firm's business, deeply embedded relationships within the firm's business network, experience (often due to their early involvement in the firm business) and extensive personal commitment (Kellermans et al. 2012; Gomez-Mejia et al. 2010). Taken together, these factors drive family managers to encourage innovation processes aimed at ensuring the firm's value creation and technological development. As a further specification of this general result, the empirical findings show that family managers have a positive influence on both the *value* and the *basicness* of firm innovation. In other words, besides favoring innovative activities that contribute to generate impactful innovations that yield more immediate economic value for their firms, family managers also seem to recognize the importance of exploring more uncertain yet potentially more remarkable inventive paths, consistent with the specificities of the industry context. Once again, this finding, combined with the lack of significant family ownership effects, seems to be in line with the existing literature that suggests distinguishing between active and passive family involvement (Maury 2006) and argues that only when the family is engaged in the firm's business life beyond mere ownership rights, a positive effect of the family-specific resources and capabilities materializes.

A third set of results relates to the role of *family supervision*, which captures the family participation in the board of directors. In this respect, the findings suggest that family supervision has a positive effect on the *scale* of innovation, but a negative influence on its *basicness*. This combination of a positive effect on the quantitative dimension of firm innovation performance and a negative effect on a specific qualitative dimension of firm innovation performance is very intriguing—particularly if related to the specificities of the study empirical context—since it seems to echo recent advances in the behavioral agency theory of family firms (e.g., Chrisman and Patel 2012). High-technology intensive industries make innovation an essential ingredient of firm survival. Hence, it could be argued that, in light of their control and advisory authority (Bammens et al. 2011), family-influenced boards of directors encourage the creation of large patent portfolios, which may support the firm's ability to smoothly adapt to technological changes in the external environment, thus avoiding the risk of organizational failure; simultaneously, they are also likely to dissuade the firm's involvement in innovation projects that embed a higher risk profile, such as those that make an extensive use of fundamental knowledge. In other words, while these results are seemingly contradictory, in reality, they are consistent with novel perspectives that build upon the behavioral agency model leveraging the construct of socio-emotional wealth (SEW) and mixed gambles (Chrisman and Patel 2012; Gomez-Mejia et al. 2014). According to this line of research, family decisions are driven by an aversion to the loss of the SEW, rather than to risk in general. This book's findings on family supervision are consistent with this idea. On the one hand, family-influenced boards of directors promote the scale of a firm's innovation as a way to preserve the firm continuity, in spite of the risk profile embedded in innovative activities; in fact, in high-technology contexts, it is the resistance to change, rather than the continuous development of innovations, that carries the highest chances of failure and, in turn, the greatest risk of SEW loss. On the other hand, they also discourage overly explorative innovative projects: These have a greater potential to generate significant breakthroughs but, at the same time, expose loss-averse family firms to higher chances of catastrophic consequences in case of project failure. Finally, the empirical analysis also seems to point to a positive effect of *family supervision* on the *value* of firm innovation. This result has been emphasized less in this book, because a set of robustness checks questions the stability of its significance. Yet, it is in line with the predictions of Matzler et al. (2015)—who also report a positive yet non-significant effect of *family supervision* on a firm's patent citations—and seems to suggest that the family members in the firm's board of directors tend to support the firm's involvement in innovative projects aimed at the creation of commercial value. Hence, while they are likely to warn against innovation projects that are perceived as excessively risky and distant from the market, such as those that involve an extensive leverage of basic knowledge, they can be expected to promote innovative activities that likely yield immediate commercial results, to the benefit of the company's viability. On the whole, these findings lend support to the behavioral agency model of family firms, suggesting that family board members could simultaneously be risk willing (e.g., when they promote innovative activities, which are inherently risky, to pursue the *scale* and *value* of innovation in recognition of their importance for the firm long-term survival in a high-technology setting) and risk averse (i.e., when they discourage the *basicness* of innovation, which would excessively boost the firm's risk profile to strive for innovation goals that are not essential to the firm's continuity).

Taken together, the aforementioned findings seem to suggest that different theoretical perspectives should be used to explain how the several dimensions of a family involvement in a firm's business affect the performance of technological innovation. Specifically, agency and resource-based perspectives complement each other to explain the role of family management, while the behavioral agency model seems to prevail as the appropriate lens to interpret the effect of the presence of family members in the firm's board of directors.

6.2   Contributions

This book offers several contributions to the nascent literature on technological innovation in family firms. First, unlike the majority of existing studies that have focused on innovation inputs, this work explores the family's influence on innovation outputs. This is important because, while innovation inputs are merely determined by a managerial decision, innovation outputs are also influenced by the firm's overall innovation orientation and by the set of resources, competences, and processes the firm can mobilize in support of innovative activities (Matzler et al. 2015). The findings of this book, thus, depict not only the family influence on the managerial decision to invest in innovation, but also a more general impact it exerts on the organizational inclination toward certain types of innovative activities.

Second, building on previous family business and technological innovation literature, this book unpacks both the family construct and the technological performance construct, in recognition of the multidimensionality and complexity they incorporate. Specifically, while the influence of the family is analyzed along the dimensions of ownership, management, and governance, the innovation performance is explored in terms of scale, value, basicness, and technological scope. This approach allows to identify a number of relationships linking specific governance and innovation variables, which more aggregate definitions of these constructs would not have unveiled. Taken together, the empirical findings demonstrate that the architecture of the influences of the family on a firm's innovation performance is very complex and articulated because, as envisaged by a recent conceptual piece by De Massis et al. (2015), different dimensions of the family construct may heterogeneously affect different facets of firm innovation performance. Hence, not only do the results confirm the importance of adopting a fine-grained approach to the study of innovation performance in family firms, but they also suggest that only a flexible and comprehensive account of the heterogeneity embedded in both the family and the innovation performance concepts may allow to disentangle the full set of relationships linking them. For instance, *family supervision* has a positive effect on the *scale* and the *value* of firm innovation, but it also negatively affects its *basicness*. Empirically, focusing exclusively on either one of these dimensions of innovation performance would have prevented to capture this puzzling effect. Acknowledging that specific dimensions of the family influence may have contrasting effects on different traits of firm innovation performance may provide a rationale for understanding why existing empirical evidence on this phenomenon is often ambiguous and should warn against the generalization of findings that involve individual dimensions of a firm's innovation performance (e.g., Duran et al. 2016), since family might act as a double-edged sword that fosters innovation scale but hampers innovation quality. Theoretically, the approach of this book drives to conclude that agency theory, behavioral agency model, and resource-based views—taken individually—cannot fully explain the multifaceted relationship between family and innovation output; rather, they should be used in combination to provide a more thorough understanding of the complex set of influences that the family may exert on firm innovation performance. These do not only arise from the alignment among the family ownership and other dimensions of a firm's leadership and governance, but may also depend on the resources, incentives, and behavioral approaches of the family members who, at various levels and in different positions, are involved in the firm's business. Hence, this book suggests that a multitheoretical approach is needed to comprehensively grasp the complex phenomenon of how family firms perform within their technological innovation activities.

This study also offers insights to the stream of literature aiming to disentangle the role of the board of director in family firms (e.g., Bammens et al. 2011). Indeed, the positive relation between family supervision and the scale of innovation can be interpreted as the result of the family focus on non-financial goals. The desire to preserve the continuity of family command, the family harmony, and the social status (e.g., Gomez-Mejia et al. 2007; De Massis et al. 2015) may thus have positive implications for a firm's competitive position (in this study, a higher propensity to innovate), in contrast to prior works that identified in the extraction of private benefits the consequence of non-financial goals in family firms (Anderson and Reeb 2004).

Finally, this book contributes to the empirical literature on innovation in family firms by focusing on a setting, namely the pharmaceutical industry, which is relatively novel to the research questions explored in this book (Gomez-Mejia et al. 2014). The characteristics of this context, and specifically its high-technology intensity, make it a particularly intriguing setting for the investigation into the relationships between family influence and firm innovation performance. In this type of contexts, family firm's alleged aversion to (inherently risky) innovative activities conflicts with the industry-specific structural need to innovate. Therefore, as highlighted by Gomez-Mejia et al. (2014), the salience of family firms' innovative behavior is greater because the high-technology intensity of the industry setting exaggerates the potential gains and losses arising from firms' innovation decisions. In addition to the industry-specific characteristics, which certainly affect the prominence of innovation for firm survival and, in turn, contribute to determine how the family strategically seeks to influence the firm's innovation performance, the investigation of the Italian pharmaceutical industry conducted in this study seems to suggest that institutional and regulatory frameworks are also critical to the relationships linking family business and technological innovation. For instance, interviews with industry experts and non-family managers have systematically emphasized that Italian family firms' involvement into top-notch innovative activities seems to be constrained by a number of inadequate institutional conditions and regulatory barriers at the country level (as highlighted in Chap. 4), which significantly lower family firms' incentives to become truly innovative and compete with international rivals striving to push the technological frontier. In other words, it seems that family firms could be less able than non-family firms to react to adverse influences arising from the external environment. This could be explained in light of their inclination to path-dependency and continuity, which might drive them to avoid any notable change in their way of dealing with innovative activities, rather than seeking a decisive upgrade that might instigate radical modifications of the status quo. Given its single-country approach and the limited sample size, this book is unable to isolate the role of institutional and regulatory variables, yet future studies should account for the interaction among these multilevel factors, as represented in the analytical framework proposed in Fig. 6.1.

6.2.1 Limitations and Future Research Agenda

The aforementioned findings and contributions should be considered in light of the study's limitations. First, the focus on the Italian pharmaceutical industry and the panel structure of the data have significantly constrained the amount of information available for analysis. For instance, the empirical investigation does not control for R&D intensity of the sample firms, since—as highlighted by previous studies (Zona et al. 2013)—these data are not accessible through the database used in this book. While the use of firm fixed-effects specifications either as main models or as a robustness check reduces the seriousness of this problem, future studies should leverage more comprehensive datasets in order to better isolate the family–innovation relationships. The focus on Italy also limits the generalizability of the findings and, at the same time, reduces the scope for investigating the role of different institutional and regulatory frameworks. Hence, a multicountry study is required to disentangle the effect of institutional factors that may affect family firms' innovative behavior and performance (e.g., Duran et al. 2016). Likewise, the empirical analysis conducted in this book should be extended to other sectors featuring both similar and different levels of technology intensity, in order to ascertain to which extent the results are driven by this industry characteristics. Moreover, while the

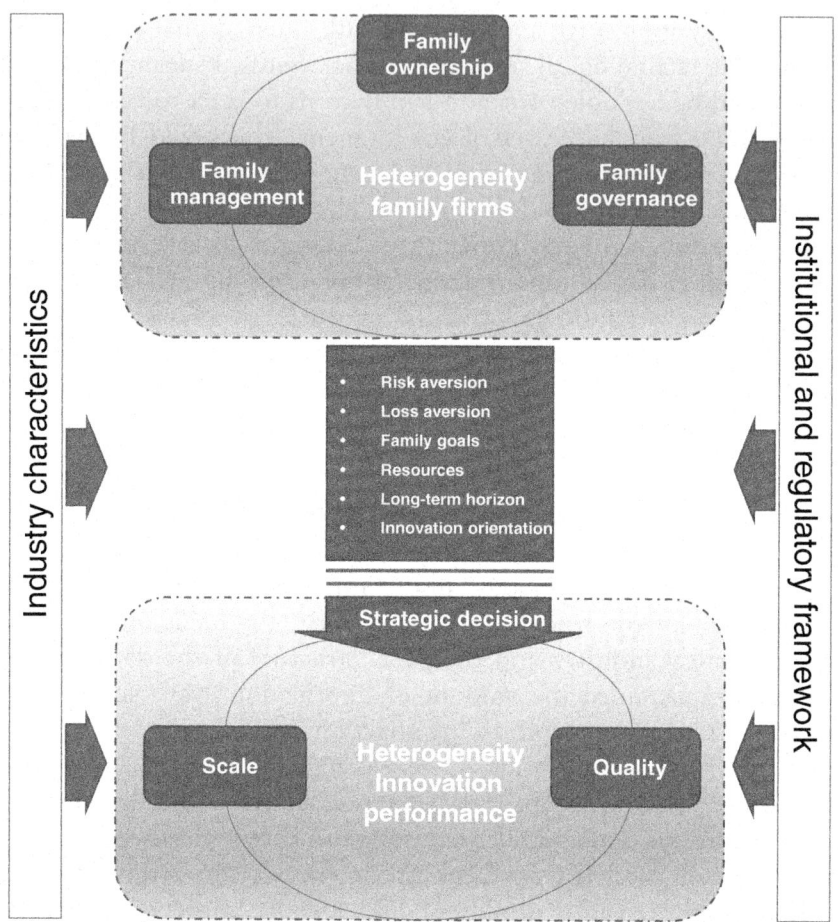

Fig. 6.1 Analytical framework

analysis has benefited from the qualitative insights provided by a number of industry experts who helped to frame and interpret the results of
the quantitative analysis, researchers should make a more extensive use
of qualitative research methods that allow to identify more accurately
the processes through which the relationships observed in this study
emerge, as well as to better capture the heterogeneity of family firms.

Another relevant limitation is the issue of endogeneity. This study controls for potential endogeneity problems associated with family-managed firms, estimating a set of two-step treatment effects models, in which the dependent variable of the first stage is family CEO, while the second stage relies on the predicted value of family CEO to estimate the various innovation performance variables (Miller et al. 2013). However, several additional sources of endogeneity are not considered in this book due to data constraint. A potential sample selection bias may occur, since firms filing patent applications may systematically differ from those not applying for patents. Hence, future work should consider a two-stage approach to predict the decision to file for patent application and alleviate this potential bias. Moreover, it is also possible that innovation inputs could be endogenous with regard to innovation outputs (Matzler et al. 2015). Also, in this case, a two-stage approach could be used to control for systematic differences in firms' innovative activities. Finally, although the use of fixed-effects specification reduces the possibility of another source of endogeneity, i.e., omitted variables, there might be other variables at institutional or at individual levels that affect innovation outcomes (e.g., Lodh et al. 2014).

Third, there may be other governance characteristics that could influence innovation performance. In this regard, it could be particularly helpful to investigate around the role of the founder. Previous studies have shown that founder-managed firms invest more on innovative activities (Block 2012) and are more likely to produce radical and exploratory innovations (Block et al. 2013). However, founder-led firms seem also to exhibit a lower level of innovation output (Duran et al. 2016). Future works might reconcile these puzzling evidence, by investigating on how founder-led firms differ from other family firms with the regard to *innovation scale, basicness,* and *value.* Similarly, the involvement of later generations in either the governance or the management of the firms opens up new possible lines of inquiry. Moreover, future works could build upon these findings and investigate the interaction among the three dimensions of the family influence on a firm's business, i.e., ownership, management, and governance. In this vein, it could be interesting to understand whether different configurations of family involvement could shape firm innovation performance (Nordqvist et al. 2014).

To identify family board members, the criterion of surname affinity is used in this work. However, it is hard to consider non-family board members a homogenous group. Indeed, most of them could have substantial linkages with either the firm or the family (e.g., consultancy relationship) limiting their ability to monitor family firms' opportunistic behavior (e.g., Dalton et al. 2007). Future work should recognize both theoretically and empirically any potential source of diversity among board members that could shape both the monitoring and the advising role of board of directors. In this vein, since the top management team represents the most significant intersection between family and business (Gersick et al. 1999), its composition also represents a potential driver of innovation output. For example, Binacci et al. (2016) have shown how diversity inside the non-family component of the top management influences family firm performance. Building on these findings, future work could shed light on how the demographic characteristics of top management teams shape innovation activities and contribute to the dialectic between (SEW-protecting) family and (business-protecting) non-family managers with the aim to find the proper balance between emotional and economic objectives.

As most of the study on the relation between family and innovation, this research is unable to disentangle the processes linking these variables. To gain deeper insights into the innovation performance of family firms, future studies should adopt a qualitative approach to investigate how innovative activities in family firms are conducted, offering a more fine-grained understanding on how intra-family dynamics, such as trust, conflicts, rivalries (e.g., Bammens et al. 2011), shape innovative decisions and outcomes. For instance, interviews with industry experts and family managers have confirmed how heterogenous the family member involvement in innovation decisions is. A personal interview with the board Chair of an Italian pharmaceutical company clearly reveals that in this company the board of directors, characterized by a strong presence of family members, has the final decision on the directions of the firm innovative activities, although the R&D director is a non-family member recruited to enhance the firm technological performance. Other Italian companies are instead characterized by a direct involvement of family members in the R&D management. Therefore, future works should further clarify the puzzling role of family involvement in the management of firm innovation.

6.2.2 Managerial Implications

These results are of interest for both managers and policy makers. Since the empirical findings reveal a positive impact of family CEO on qualitative dimensions of a firm innovation, but do not report any significant effect of family ownership, firms should be aware that only an active involvement of the family in the firm business can help to enhance the qualitative profile of firm innovation. With a direct participation in the business, the family seems able to create an innovation mentality (Zahra 2005) that fosters the *value* and the *basicness* of firm innovation. This result is extremely relevant for policy makers as well, particularly in light of the high-technology intensity of this study setting. In fact, the development of high-technology intense industries is considered a strategic priority for many countries, but family firms are often perceived as being unfit to successfully compete in such contexts. This book shows that, precisely in high-technology intense sectors, family-managed firms tend to generate more valuable and more basic inventions. Hence, policy makers should welcome the diffusion of an organizational culture that motivates family members to involve in the management of their firms.

This book also offers a fresh perspective on the role of family board members, which should be used as inspiration by both family firms' managers and policy makers. The positive effect family board members exert on the *scale* and the *value* of firm innovation can be interpreted in light of the need to preserve the family wealth, but also in relation to the family members' endowment with firm-specific and/or industry-specific knowledge (e.g., Dalton et al. 2007). Hence, managers should be aware of this effect and take appropriate decisions, for instance in terms of compensation and incentive plans.

On the other hand, the negative influence family board members exert on the basicness of innovation suggests that in countries like Italy, where the great majority of companies are family firms and the presence of family members in boards of directors is also very widespread, the industrial system's ability to generate fundamental knowledge and contribute to advancement of the technological frontier could suffer. Hence, policy makers in such countries should favor an institutional setting that helps family firms—and, specifically, family board members—rethink their relationship with the scientific world to promote more explorative innovation projects, thus alleviating the negative effect induced by their willingness to preserve the family SEW. To reach this aim, one possible solution could be promoting an institutional environment that eases the fruitful connections between firms and research organizations, such as universities and research centers, to grant family firms a more immediate access to scientific knowledge that can significantly inspires and nurtures ambitious innovative projects. Analogous initiatives to limit the negative disposition of family board members toward more basic innovation projects could be adopted within the firm itself. For instance, leveraging the common family belonging, the family management could activate a constructive debate with family board members to raise awareness on the importance of scientific knowledge as an input to the firm innovation funnel in high-technology intensive industries.

References

Anderson, R. C., & Reeb, D. M. (2004). Board composition: Balancing family influence in S&P 500 firms. *Administrative Science Quarterly, 49*, 209–237.

Bammens, Y., Voordeckers, W., & Van Gils, A. (2011). Boards of directors in family businesses: A literature review and research agenda. *International Journal of Management Reviews, 13*, 134–152.

Binacci, M., Peruffo, E., Oriani, R., & Minichilli, A. (2016). Are All non-family managers (NFMs) equal? The impact of NFM characteristics and diversity on family firm performance. *Corporate Governance: An International Review, 24*(6), 569–583.

Block, J. H. (2012). R&D investments in family and founder firms: An agency perspective. *Journal of Business Venturing, 27*(2), 248–265.

Block, J., Miller, D., Jaskiewicz, P., & Spiegel, F. (2013). Economic and technological importance of innovations in large family and founder firms: An analysis of patent data. *Family Business Review, 26*(2), 180–199.

Carney, M. (2005). Corporate governance and competitive advantage in family-controlled firms. *Entrepreneurship Theory & Practice, 29,* 249–265.

Chrisman, J. J., & Patel, P. C. (2012). Variations in R&D investments of family and nonfamily firms: Behavioral agency and myopic loss aversion perspectives. *Academy of Management Journal, 55*(4), 976–997.

Dalton, D. R., Hitt, M. A., Certo, S. T., & Dalton, C. M. (2007). Chapter 1: The fundamental agency problem and its mitigation. *Academy of Management Annals, 1*(1), 1–64.

Davis, J. H., Schoorman, F. D., & Donaldson, L. (1997). Toward a stewardship theory of management. *Academy of Management Review, 22*(1), 20–47.

De Massis, A., Di Minin, A., & Frattini, F. (2015). Family-driven innovation: Resolving the paradox in family firms. *California Management Review, 58*(1), 5–19.

Duran, P., Kammerlander, N., Van Essen, M., & Zellweger, T. (2016). Doing more with less: Innovation input and output in family firms. *Academy of Management Journal, 59*(4), 1224–1264.

Dyer, W. G. (2006). Examining the "family effect" on firm performance. *Family Business Review, 19*(4), 253–273.

Gambardella, A. (2005). Patents and the division of innovative labor. *Industrial and Corporate Change, 14*(6), 1223–1233.

Gambardella, A., Harhoff, D., & Verspagen, B. (2008). The value of European patents. *European Management Review, 5*(2), 69–84.

Gersick, K. E., Lansberg, I., Desjardins, M., & Dunn, B. (1999). Stages and transitions: Managing change in the family business. *Family Business Review, 12,* 287–297.

Gomez-Mejia, L. R., Campbell, J. T., Martin, G., Hoskisson, R. E., Makri, M., & Sirmon, D. G. (2014). Socioemotional wealth as a mixed gamble: Revisiting family firm R&D investments with the behavioral agency model. *Entrepreneurship Theory and Practice, 38*(6), 1351–1374.

Gomez-Mejia, L. R., Haynes, K. T., Nunez-Nickel, M., Jacobson, K., & Moyano-Fuentes, J. (2007). Socioemotional wealth and business risks in family controlled firms: Evidence from Spanish olive oil mills. *Administrative Science Quarterly, 52,* 106–137.

Gomez-Mejia, L. R., Makri, M., & Larraza Kintana, M. (2010). Diversification decisions in family controlled firms. *Journal of Management Studies, 47*(2), 223–252.

Habberson, T. G., & Williams, M. L. (1999). Are source-based framework for assessing the strategic advantages of family firms. *Family Business Review, 12*(1), 1–25.

Hall, B. H., Jaffe, A. B., & Trajtenberg, M. (2001). *The NBER patent citation data file: Lessons, insights and methodological tools* (No. w8498). National Bureau of Economic Research.

Jensen, M. C., & Meckling, W. H. (1976). Theory of the firm: Managerial behavior, agency costs and ownership structure. *Journal of Financial Economics, 3*(4), 305–360.

Kellermanns, F. W., Eddleston, K. A., Sarathy, R., & Murphy, F. (2012). Innovativeness in family firms: A family influence perspective. *Small Business Economics, 38*(1), 85–101.

Lee, J., & Berente, N. (2012). Digital innovation and the division of innovative labor: Digital controls in the automotive industry. *Organization Science, 23*(5), 1428–1447.

Lodh, S., Nandy, M., & Chen, J. (2014). Innovation and family ownership: Empirical evidence from India. *Corporate Governance—An International Review, 22*(1), 4–23.

Matzler, K., Veider, V., Hautz, J., & Stadler, C. (2015). The impact of family ownership, management, and governance on innovation. *Journal of Product Innovation Management, 32*(3), 319–333.

Maury, B. (2006). Family ownership and firm performance: Empirical evidence from western european corporations. *Journal of Corporate Finance, 12*(2), 321–341.

Miller, D., Minichilli, A., & Corbetta, G. (2013). Is family leadership always beneficial? *Strategic Management Journal, 34*(5), 553–571.

Nordqvist, M., Sharma, P., & Chirico, F. (2014). Family firm heterogeneity and governance: A configuration approach. *Journal of Small Business Management, 52*(2), 192–209.

Siebels, J.-F., & zu Knyphausen-Aufseß, D. (2012). A review of theory in family business research: The implications for corporate governance. *International Journal of Management Reviews, 14*, 280–304.

Trajtenberg, M. (1990). A penny for your quotes: Patent citations and the value of innovations. *The Rand Journal of Economics*, 172–187.

Trajtenberg, M., Henderson, R., & Jaffe, A. (1997). University versus corporate patents: A window on the basicness of invention. *Economics of Innovation and New Technology, 5*(1), 19–50.

Zahra, S. A. (2005). Entrepreneurial risk taking in family firms. *Family Business Review, 18*(1), 23–40.

Zona, F., Zattoni, A., & Minichilli, A. (2013). A contingency model of boards of directors and firm innovation: The moderating role of firm size. *British Journal of Management, 24*(3), 299–315.

Index

© The Editor(s) (if applicable) and the Author(s) 2017
A. Perri and E. Peruffo, *Family Business and Technological Innovation*,
DOI 10.1007/978-3-319-61596-7

CPSIA information can be obtained
at www.ICGtesting.com
Printed in the USA
LVOW13*2350021017
550905LV00013B/294/P